Chemical Process
Hazard Review

ACS SYMPOSIUM SERIES 274

Chemical Process Hazard Review

John M. Hoffmann, EDITOR
Daniel C. Maser, EDITOR
Parke-Davis Division
Warner-Lambert Company

Based on a symposium cosponsored by
the Division of Chemical Health
and Safety and the Industrial Division
of the National Safety Council
at the 187th Meeting
of the American Chemical Society,
St. Louis, Missouri,
April 8–13, 1984

American Chemical Society, Washington, D.C. 1985

Library of Congress Cataloging in Publication Data

Chemical process hazard review.
  (ACS symposium series, ISSN 0097-6156; 274)

  Includes index.

  1. Chemical processes—Safety measures—
Congresses.

  I. Hoffmann, John M., 1934-   . II. Maser, Daniel
C., 1955-   . III. American Chemical Society.
Division of Chemical Health and Safety. IV. National
Safety Council. Industrial Division. V. American
Chemical Society. Meeting (187th: 1984: St. Louis,
Mo.) VI. Series.

TP149.C487   1985   660.2′804   85-1259
ISBN 0-8412-0902-2

ACS Symposium Series

M. Joan Comstock, *Series Editor*

Advisory Board

FOREWORD

The ACS SYMPOSIUM SERIES was founded in 1974 to provide a medium for publishing symposia quickly in book form. The format of the Series parallels that of the continuing ADVANCES IN CHEMISTRY SERIES except that, in order to save time, the papers are not typeset but are reproduced as they are submitted by the authors in camera-ready form. Papers are reviewed under the supervision of the Editors with the assistance of the Series Advisory Board and are selected to maintain the integrity of the symposia; however, verbatim reproductions of previously published papers are not accepted. Both reviews and reports of research are acceptable, because symposia may embrace both types of presentation.

CONTENTS

Preface..ix

1. Chemical Process Hazard Review.....................................1
   John M. Hoffmann

2. Process Hazard Review in a Chemical Research Environment.............7
   Mary J. Hofmann

3. Hazards Evaluation in Process Development..........................17
   Daniel P. Brannegan

4. Risk Assessment Techniques for Experimentalists....................23
   David J. Van Horn

5. Hazard and Operability Study: A Flexible Technique for Process System
   Safety and Reliability Analysis....................................33
   A. Shafaghi and S. B. Gibson

6. Hazard Avoidance in the Processing of Pharmaceuticals...............41
   John R. Handley

7. Thermochemical Hazard Evaluation..................................57
   Robert C. DuVal

8. Thermal Runaway Reactions: Hazard Evaluation......................69
   Linda Van Roekel

9. The Thermochemical and Hazard Data of Chemicals: Estimation Using the
   ASTM CHETAH Program..81
   Carole A. Davies, Irving M. Kipnis, Malcolm W. Chase, and
   Dale N. Treweek

10. Kinetic and Reactor Modeling: Hazard Evaluation and Scale-up
    of a Complex Reaction..91
    Ashok Chakrabarti, Edwin C. Steiner, Craig L. Werling, and
    Mas Yoshimine

11. The Nitration of 5-Chloro-1,3-dimethyl-1H-pyrazole: Risk Assessment
    Before Pilot Plant Scale-up......................................107
    James R. Zeller

Author Index...115

Subject Index..115

PREFACE

ASSURANCES THAT NEW AND EXISTING CHEMICAL PROCESSES are conducted safely have never been more needed. Public awareness of the effects of chemical exposure has increased since the early 1970s. Although the initial focus of the Occupational Safety and Health Act of 1970 was on safety, clearly the emphasis now is on health. People at all levels of society are concerned about exposure to chemicals and the possible short- and long-term effects of chemicals on human health. The effects of chemicals on the environment from past or present waste sites, accidental releases or spills, and fires and explosions are reported daily in the news media. Control of all chemical processes to avoid accidental discharges and/or upsets that lead to fires, explosions, and environmental release is essential in the laboratory, the pilot plant, and the manufacturing plant. Chemical process hazard reviews are necessary at each step in the development of a process to ensure that the process can be controlled and conducted so as to minimize the risks to personnel, property, and the environment.

The purpose of the symposium upon which this book is based was to provide a forum for the exchange of information on chemical process hazards reviews by industrial research and development chemists, chemical engineers, and safety professionals. The chapters in this text are representative of the subjects presented at the symposium and are provided to give wider dissemination and availability of this information.

We are indebted to the executive committees of the ACS Health and Safety Division and the NSC Industrial Division Chemical Section for their interest and support. A special thanks to each of the authors for their timeliness and, more important, their willingness to take time out of their schedules to share their knowledge and experience with others. Finally, we must acknowledge the Warner-Lambert Company, Parke-Davis Division, for continued support, materially and philosophically.

JOHN M. HOFFMANN
DANIEL C. MASER
Parke-Davis Division
Warner-Lambert Company
Rochester, MI 48063

November 3, 1984

Chemical Process Hazard Review

JOHN M. HOFFMANN

Parke-Davis Division, Warner-Lambert Company, Rochester, MI 48063

The phrase Chemical Process Hazard Review has
widely varying meanings. To the professional
safety engineer, it connotes a broadly based
review of a chemical process which when conducted properly would provide assurances that
a process can be conducted "safely"; safe for
the scientists in the laboratory, the techni-
cian in the pilot plant and the chemical operators in the manufacturing plant. The developing chemist may view a chemical process hazard
review as just another hurdle to jump enroute
to getting his/her process into the pilot plant
or manufacturing plant. Chemical engineers
destined to design both the process and equipment may depend on such a review to provide
the details necessary to design the process
"safely." Environmental engineers and control
specialists consider a review process as a means
for estimating risk to environmental exposure
(air, water and ground) and a source of data
to develop compliance information. Management
needs the assurance from both line managers
and staff functions that the process can be
conducted and the hazards associated with
that process in its entirety are identified
to the extent that appropriate risk/benefit
decisions can be made. An appropriate
Chemical Process Hazard Review can and should
meet all of these needs.

Life in the chemical process development world is no longer one
shrouded in the mysteries that synthetic organic chemistry can present. The basic researcher does not have the luxury of living in

0097–6156/85/0274–0001$06.00/0
© 1985 American Chemical Society

an isolated laboratory and day after day producing new chemical
entities through ever increasing complex chemistry without regard
to the ultimate process in the pilot plant. The developmental
chemist's concern is not just defining a workable, cost-effective
process but one which can be done "safely" and economically. The
chemical engineer designing a plant scale process also has added
a number of new concerns which must be considered before committing
to a plant installation.

Consider for a moment just the changes in vocabulary during
the past twenty years among these professions. Chemists and
chemical engineers long known for their seemingly capricious use
of acronyms and peculiar jargon are faced with incorporating and
understanding those invented and used by others. OSHA, PEL, NIOSH,
RTECS, HMTA, FWPCA, CPSA, IARC, FEPCA and SF are just a few. With
this seemingly overwhelming burden of regulatory requirements, life
and business must go on and they are. A small part of continuing
life and business has been and is in the Chemical Process Hazard
Review.

Notwithstanding the complexities considered above, chemical
processes must be developed in a way which people, the environment
and property are protected. The chemical reaction process itself
presents its own peculiar evaluation needs which impact upon
occupational safety and health, environmental protection and property preservation and conservation. In the progression of a process
from the research laboratory to the development laboratory to the
pilot plant and eventually to manufacturing, the objectives for the
process review must be:
1.   The ability to carry out the desired process producing the
desired products, profitably;
2.   Development of data which can characterize the reaction, including side or minor products, and the conditions under which they
may be produced.
3.   Development of reaction conditions (temperatures, pressure,
concentrations and equipment) at which the process can be
conducted "safely."
4.   Identification of conditions and events (other than normal)
which can lead to hazardous conditions or products.

Most major chemical and pharmaceutical companies today have
developed systematic methods of evaluating new (and in many cases,
old) processes and materials for the hazards attendant to their
manufacture. The degree of urgency in establishing a chemical process hazard analysis function has often been dictated by some untoward event (usually within the company). It is to the prediction
and control or elimination of unplanned reaction events to which the
chemical process hazard review must address itself.

Broadly, the review process can be segregated into: an understanding of the chemistry, desired and possible; a thorough literature review of related chemical reactions and processes; theroreti-
cal calculations; and design and conduct of experiments to confirm
hypotheses and/or gather additional data.

Process Chemistry

In understanding the desired and possible chemistry the process at each step should be categorized as to type of reaction and a balanced chemical equation developed. The use of a broad categorization such as given in Table I is frequently used as a starting point. All probable side products should be identified and their potential role(s) postulated. Rate determining steps in multi-step or sequential processes should be so labeled and their import to the process identified.

Table I.   Chemical Reaction Energy Categorization

Process	Energy	Chemical Hazard Potential
Oxidation	Highly Exothermic Equilibrium Favored	High
Nitration	Exothermic Potential oxidation	High
Reductions	Low	Low
Halogenations	Highly Exothermic chain reaction for Chlorine & Fluorine	High
Sulfonations	Moderately exothermic	Low
Hydrolysis	Mildly exothermic	Low
Polymerization	Can be highly exothermic	Moderate to high
Condensations	Moderately exothermic	Low to moderate
Hydrogenation	Mild to moderately exothermic easily controlled	Moderate to low
Alkylation	Mildly exothermic side reactions generally a problem	Low
Organo Metalics	Highly exothermic	High
Amination	Moderately exothermic	Low

Identification and quantification of the desired reaction conditions, particularly temperature and concentration, are necessary to evaluate what may happen if these conditions are not met. This is particularly true where equilibrium considerations are a significant factor in a rate determining step between or among competing reactions. Where multiple products are possible, temperature variations will often significantly alter the ratios of these products. If one of these is unstable or more toxic, this could lead to more stringent temperature control requirements in the process and equipment design.

Identifying the "type" of reaction can be useful in broadly categorizing the overall potential hazards as well as aiding in the literature search. Many literature sources refer to reactions not only by their historical name but also by their type. If a reaction step in a process can be identified in this manner, it may aid in the literature search and in keying various reviewers memories. Types of reactions such as Fridel-Crafts, Grignard, Meerwin-Ponndorf, Cannizzaro, Clemenson. Wolff-Kishner, Hofmann Degredation, Beckman Rearrangement, Simmons-Smith, Piels-Alder, Wurtz-Fitig, etc. may not be household names to everyone, but they are to many organic chemists. More importantly, literature sources can be searched for these classifications to give an overall perspective to a reaction under study.

Literature

Not enough can be said about literature reviews when you consider how many chemists there are in the world today (128,000 current members in the American Chemical Society alone) and when you consider how many there have been from the days of Priestly, the number is awesome. Many of these chemists as experimentalists have tried mixing a little of everything with anything to see what would happen. Often something dramatic did happen. Much of this information has been recorded in the literature albeit a bit hard to find at times. Frequently, chemical accidents in the laboratory are reported in C&E News in the letters to the editor. Invariably, after publishing the untoward event in this manner other readers will respond with reports of similar occurences that they were involved in or knew about. And others will report a like event recorded in the literature in years past. Table II lists a few starting points for literature reviews. As is often the case in literature searches, one key article can often provide the route back to previous authors and reviewers who have worked in that particular area of chemistry.

Table II. Literature Review

---

o   L. Bretherick - Hand Book of Reactive Chemical Hazards
o   NFPA 491 M & 49
o   Factory Mutual Data Sheets 7-19, 7-23
o   Kirk-Othmer
o   CMA (MCA) Accident Case Histories Volumes 1-4
o   Open Literature Via Chemical Abstracts
o   Data Bases Such As:
       Med Lars, Tox Line
       Toxicology Data Bank
       and D D C.

Theoretical Calculations

Theoretical calculations can be made regarding the energy which is available from a reaction. The use of chemical thermodynamic tables, tables of heats of formation and known heats of reaction can lead to estimations of the energy available. Coupling this with reaction conditions (temperature particularly) and gross rates observed on bench-scale experiments can produce meaningful parameters translatable to scale-up.

Calculations of oxygen balance either for the total reaction mixture or for individual products is a simple yet very effective first-order evaluation of a process or material, particularly nitrations and nitro-containing compounds. Most authors recommend determining the ratio of oxygen available within the system to theoretical oxygen requirements necessary to oxidize all carbon to carbon dioxide, all hydrogen to water, and nitrogen reduced (or in some cases, oxidized) to elemental nitrogen. If other oxidizable species are present, such as sulfur or alkali/alkaline earth metals, these would also be presumed to be oxidized to their oxides in the most stable oxidation state.

The CHETAH Program from ASTM Committee E-27 is also quite useful for theoretical calculations of enthalpy (decomposition, oxidation or combustion), oxygen balance, and potential energy release. Recent modifications of the program and updating of the data base make it even more useful.

Testing

Evaluation of a material or process by analytical tests and experiments is, of course, a must when history (literature), theory and paper chemistry are insufficient to characterize the process or answer the significant "what ifs." Table III lists some of the more common useful methods.

Table III.   Testing Methods

DTA	Differential Thermal Analysis
DSC	Differential Scanning Calorimetry
TGA	Thermo Gravimetric Analysis
ARC	Accelerating Rate Calorimetry
BSC	Bench Scale, Heat Flow Calorimetry
SEDEX	Sensitive Detector of Exothermic Processes
Others	Oven Tests, Dewar Tests, Hot Plate Tests, etc.

The ultimate purpose of these types of tests is to evaluate two similar (in results) but different occurrences. These are runaway chemical reactions and exothermic chemical decompositions. The first may actually just be a desired reaction out of control while the second is an undesired reaction out of control. Among the purposes which analytical tests serve are the determination of the "onset" of exothermic (endothermic) decomposition. While frequently a specific temperature is cited for such "onsets," one must remember that this temperature is highly dependent on instrument sensitivity, degree of adiabaticity and time-temperature history. It should be stated that tests results are accurate only for the exact conditions under which they were run. Physical factors such as density and geometry can also influence test data. In theory, reaction rates are not a step function but are continuous. A reaction rate for a process is not zero below a given "onset"

temperature, but is merely a smaller number which instrumentation
is not sufficiently sensitive to measure. In practice, however,
established temperature onsets for process reactions will usually
allow a sufficient margin of safety, provided that adequate cooling
capability or inherent heat sinks are sufficient to remove heat
energy in excess of that which can be generated. For highly exothermic reactions or reactions with low activation energies, it
may be necessary to modify the process. Continuous or semi-batch
versus batch is one way in which control may be maintained within
engineering limits.

Summary

The challenge that faces chemical researchers, development chemists
and engineers, manufacturing chemists and engineers and the staff
functional professionals associated with chemical processes is
formidible. It is one that requires more and better information
about chemical processes to meet today's regulatory demands and
both public and private expectations for the chemical and pharmaceutical industries. Organized and systematic chemical process
hazard reviews are necessary to meet these demands.
       The attempt in organizing the Symposium on Chemical Process
Hazard Review at the Spring 1984 ACS national meeting in St. Louis,
was to present papers on the review procedure, some of the thermochemical evaluation techniques and the application of both of
these to actual processes. Some of the papers presented at the
symposium have been collected and are published here to further
exchange information related to conducting chemical processes
"safely."

Literature Cited

1.   Bodurtha, Frank T., Industrial Explosion Prevention and Protection, McGraw-Hill 1980
2.   Fawcett, H.H. and W.S. Wood, Safety and Accident Prevention in
Chemical Operations, Interscience Wiley 1982
3.   Bretherick, L., Handbook of Reactive Chemical Hazards, Second
Edition Butterworths and Co. 1979
4.   Meyer, Eugene, Chemistry of Hazardous Materials, Prentice-Hall
1977
5.   Seaton, W.H., Freedman, E., Treweek, D.N., CHETAH - The ASTM
Chemical Thermodynamic and Energy Release Evaluation Program, ASTM
Philadelphia
6.   Lothrup, W.C. and R. Hendrick, "The Relationship Between Per-
formance and Constitution of Pure Organic Explosive Compounds,"
Chem Rev, vol 44 no. 3 pp 419-445, June 1949
7.   Manufacturing Chemist's Association, Guide for Safety in the
Chemical Laboratory, 2nd ed. Van Nostrand Reinhold, New York 1972
8.   Skoog, D.A. and D.M. West, Principals of Instrumental Analysis
2nd ed., Saunders College, Philadelphia 1980
9.   Cardillo, P., Combined Use of Instruments for Determining Chemical Safety, Cromache Chimivca, 69, 1982
10.   Fenlon, W.J., A Comparison of ARC and Other Thermal Stability
Tests, AICHE Loss Prevention Symposium Aug. 1983

RECEIVED November 3, 1984

Process Hazard Review in a Chemical Research Environment

MARY J. HOFMANN

Experimental Station, E. I. du Pont de Nemours and Company, Wilmington, DE 19898

The objectives of the Process Hazards Review program at the Du Pont Experimental Station are reviewed. The scope, organization, format, review method, final report, and frequency are discussed as they apply to research projects. The concept of tailoring the review method to the degree of hazard involved is explained and an example of a Process Hazards Review and a Pre-Startup Review/Process Hazards Audit are given.

It has long been recognized that accidents rarely result from unforeseeable hazards. They also rarely come about from "acts of God". What we have come to realize is that accidents really result from a failure to define and control <u>known</u> hazards that exist due to the equipment, the chemicals, the chemistry, or the people involved.

In the Du Pont Company as a whole, a vigorous program of process hazards management, of which Process Hazards Reviews (PHR's) are but one element, was instituted and has been recommended by our Corporate Safety & Fire Protection Division as far back as 1966 <u>(1)</u>.

The program, of course, has been widely used at manufacturing sites, but its rigorous application to the research environment in DuPont has been fairly recent - since 1979 <u>(3)</u>. In a number of ways, it is still evolving; such as how often should a PHR be held or when is one needed?

Research at the Du Pont Company's Experimental Station encompasses virtually all fields of science - physics, chemistry, biochemistry, and engineering. The scope of experimentation will range from the micro-level, to the semiworks-level, using from milliliter quantities to a drum lot daily. The frequency of experimentation may range from a one-time run in a hood to a round-the-clock operation in a barricade. A Process Hazards Review program must encompass all these possibilities.

0097-6156/85/0274-0007$06.00/0
© 1985 American Chemical Society

Objectives

The ultimate objectives of the Du Pont Company's Process Hazards Reviews, and Pre-Startup Reviews/Process Hazards Audits, are to:

● eliminate injuries, and
● minimize property and environmental damage resulting from the process hazards.

This is done by:

● Identifying process and equipment hazards which could cause serious injuries, explosions, fires, or toxic material releases. These hazards may have been previously unrecognized; or they may have been recognized and tolerated but avoided by skilled or experienced employees.

● Evaluating the size or impact of the hazards, the potential for injury to personnel and property loss, and the frequency of occurrence,

● Developing recommendations to eliminate or control the hazards, and

● Implementing the recommendations.

Definitions

The terms Process Hazards Review and Pre-Startup Review/Process Hazards Audit have been mentioned and are defined.

Process Hazards Reviews comprise formal committee meetings where hours are spent intensively examining, by one of the methods described later, a chemical reaction or process, with a report, documentation, and follow-up. Pre-Startup Reviews/Process Hazards Audits are no less intensive, but the time spent is less, because the complexity of the process or equipment being examined is less. Reports, documentation, and follow-up are also a part of the Pre-Startup Review/Process Hazards Audit. An equipment acceptance safety inspection would be considered a Process Hazards Audit.

Scope

It is necessary to have an intensive and yet systematic examination of the process or the equipment for hazardous exposures to personnel and to property. This should be held from both a theoretical and a practical view. "What if?" situations or those not readily apparent, such as impurities in reactants, the materials of construction, or the suitability of control devices need to be emphasized.

We have made Process Hazards Reviews and Audits distinct from incident or accident investigations, although either of these may bring home the need to hold one, as may an area safety audit or survey.

An example where there was a need for a Process Hazards Review occurred recently in investigating an incident in which there was a small explosion in an oxygen supply connection to a high pressure reactor. During the investigation, it was brought out that the oxygen supply system was installed after the original Process Hazards Review was held, and that this new oxygen system had never been intensively reviewed.

Review Needed

The most difficult decision in the research environment is when to conduct a Process Hazards Review. A PHR certainly need not be held for a laboratory-scale experiment conducted in a chemical fume hood following a documented procedure - this is the one end of the spectrum. At the other end, a PHR must be held on a semiworks operation, that will be running around the clock, involving drum quantities of materials. It is in the scale in between that a decision is more difficult. The Du Pont Experimental Station has set up certain guidelines. PHR's must be held on:

● All new capital projects when specified in the project write-up. These capital projects could cover, for example, the purchase and installation of a piece of analytical equipment such as an electron microscope, or the renovation and equipping of a new polymer testing laboratory.

● All Class IV lasers. These are high power lasers having 5×10^{-1} watts of power or greater. PHR's are required because of the control measures, such as interlocks and signs; and the health hazards that exist when these lasers are in use.

Process Hazards Reviews are strongly recommended in these cases:

● New or revised operations in a semiworks area,

● Laboratory reactions that may be potentially explosive because of the reactants or products,

● Laboratory reactions using chemicals that are:

- highly toxic
- radioactive
- carcinogenic

● Laboratory operations on a large scale such as those using 22-liter flasks for reactions, isolations, purifications, etc.,

● Laboratory reactions that will be running around the clock or for more than the normal eight-hour work day, and

● Laboratory operations where standard glassware or plastic will be under pressure.

These guidelines are not meant to be all inclusive and there are cases where a combination of less hazardous conditions can create a need for a Process Hazards Review or a Pre-Startup Review/Process Hazards Audit.

Types of Reviews

The next decision to be made after the need for a PHR has been established is what type of review to hold. A look at the various types and a description of each will be helpful.

● "What if?"

The "What if?" is designed for relatively uncomplicated processes. At each step in the process or reaction "What if?" questions are asked and the answers are considered in evaluating the effects of failures of components or errors in the procedure (2).

● Checklist

For slightly more complex processes, the checklist method provides a more organized approach (2). This is accomplished by the use of lists of words or phrases that will stimulate questions concerning the subject. For example, the phrase Personnel Protection should lead to questions relating to the adequacy of ventilation and toxicity of the chemicals used. There are a number of checklists available in Du Pont, each applicable to the site or department for which it was written. Assignments of certain aspects of the project under review can be made to committee members who have the greatest expertise in that area.

● Failure Mode and Effect Analysis (FM&E)

When analysis is needed of a small portion of a large process or of an item of equipment, such as a reactor, the Failure Mode and Effect method can be used (2). While this method may not evaluate operating procedure errors or omissions, or the possibility or probability of operator error, it does assess the consequences of component failures on the process. This type of analysis has been used infrequently at the Experimental Station, and then most often in a somewhat modified form.

● Hazard and Operability (HAZOP) Study

In this method, every part of a process is examined to

discover how deviations from the intended design can occur and how these deviations can cause hazards. No HAZOP studies have been performed at the Experimental Station because the other methods have served our operations successfully.

- ## Fault Tree Analysis (FTA)

Finally the most rigorous method is the Fault Tree Analysis. In this method, a specific undesired process event such as an explosion is postulated and placed at the top of a tree, from which branches representing all possible precursor events or causes are extended. When basic causes are reached, failure rates are estimated or obtained. While much has been written about Fault Tree Analysis, this method has not been used at our research site because less rigorous methods are more suited to our rapidly changing research environment. It has been extensively used at our production sites, however.

Selection of Type of Review

The thinking and decision making used at the Experimental Station are based on the information in Table I. In this table, an evaluation of the complexity of the process as it relates to the scale of the operation is made. Then a PHR method is selected, using in ascending order of intensity, the "What if?", the Checklist, the Failure Mode & Effect, and the Fault Tree.

TABLE I. PHR SELECTION METHOD

Scale	Batch Process		Continuous Process	
	Lab/SW	Service	Lab/SW	Service
Exploratory Research	What if?	What if?	What if?	What if?
Research	What if?	What if?	What if?	Checklist
Scale-up (lbs.)	Checklist	Checklist	Checklist	FM&E
Process Development Start-up/Shutdown	FM&E	FM&E	FM&E	FTA
Freestanding Purchased Equipment	What if? or Checklist		What if? or Checklist	

What if? ⎤
Checklist ⎥— Level
FM&E ⎥ of PHR
FTA ⎦ method

As can be seen, in batch operations, the "What if?" method is most commonly used, with the Checklist and Failure Mode & Effect method used in larger, more complex operations. The Experimental Station Service organization, using its own personnel, carries out various experiments for researchers; and at the batch level, conducts essentially the same type of PHR as a laboratory or semiworks does. Falling somewhat outside of the scale concept is the category of freestanding purchased equipment, such as instruments, where a "What if?" or Checklist analysis is conducted. In a continuous operation the "What if?" or Checklist method will also be conducted to a great extent, but a Failure Mode & Effect analysis would be used in a continuous process with start-ups and shutdowns.

Process Hazards Review Committee

After the review method is chosen, the responsibility for holding the PHR or Pre-Startup Review falls to the line organization who contact the Process Hazards Review Committee. At the Experimental Station, each resident Laboratory or Department has a committee appropriate to its needs. In each case, the committee chairman is in a management position and can command the resources necessary to review the process or equipment successfully. It is the chairman's responsibility to assure that the review is intensive and covers all aspects of the process or equipment. The chairman also serves as the Laboratory's liaison with the Site Process Hazards Management Committee.

The committee itself consists of, as well as the chairman, the designer or user of the process or equipment, who arranges the meeting time and place and prepares the necessary documents. The designer or user also acts as the secretary for the review. Others on the committee include:

● The Safety Engineer from the Site Safety Office with liaison responsibility for the resident Laboratory in which the process or equipment is located,

● A technical person not connected with the process or equipment,

● A design engineer, if one was involved,

● An engineering maintenance supervisor.

This group forms the nucleus and is considered a minimum. Others with special skills may need to be involved for some reviews, such as the Site Industrial Hygienist, the Site Radiological Safety Officer, an instrumentation engineer, or a person experienced with explosion hazards.

Process Hazards Review Agenda

A typical agenda for a Process Hazards Review generally follows these lines:

●   Preferably, at least a week prior to the scheduled meeting time, the user of the process or equipment sends to each committee member for review, a document containing:

- A general statement about the purpose of the process or equipment;
- The process chemistry, such as reaction rates, cooling rates, side reactions, temperatures, and pressure;
- The process material or equipment hazards, such as toxicity, flammability, electrical, and mechanical hazards;
- The location of safety equipment;
- The procedure for emergency shutdown;
- The process flow diagram;
- A description of the equipment, including, for example, pressure and temperature ratings and controls, construction, capacity, the relief devices, and instrumentation;
- The operating instructions, including the safety precautions to be taken and protective clothing to be worn at each step or series of steps;
- The waste disposal and spill control procedures and other environmental considerations; and
- An appendix containing material safety data sheets on the chemicals used in the process or equipment and any other information of a helpful nature.

● At the scheduled meeting, the complete package of information is reviewed and discussed. This is the time the "What if?" questions are asked, or the Checklist used. If a Failure Mode & Effect analysis is used, this information also will have been provided and discussed.

● Following the discussion, the committee makes a field audit, checking the suitability and placement of equipment, the impact of the process on the area, and the impact of the area on the process.

On complex processes, the operation is generally subdivided into logical or manageable units and separate PHR's are conducted on each unit.

In almost all cases, the person responsible for the process or equipment being reviewed writes the final report which includes:

● The hazards identified, whether existing or potential,

● The committee's recommendations for corrective actions,

● The persons responsible for the corrective actions and dates for completion, actions and dates for completion, and

• A summary of all questions raised and resolved so that future PHR's on the same process will not "plow the same ground again".

This written final report is then sent to all committee members, appropriate members of management, and generally to the central file of the research laboratory in which the PHR originated.

Process Hazards Review Frequency

We recommend that re-reviews be held whenever a substantial change is made in existing operations. Changes such as increased pressures and/or temperatures, or different reactants signal the need for a re-review. Changes to fixed equipment should also be evaluated, and a decision made as to whether a re-review is necessary.

Examples

The first example is of a Process Hazards Review conducted on a flow reactor designed to test the activity of heterogeneous catalysts in the reaction of mixtures of hydrogen, carbon monoxide, and/or carbon dioxide. This reactor system had previously been reviewed and the Review this time was to assess the hazards of the addition of a vaporizer and liquid pump to the system.

A committee was assembled and consisted of:

• the Department Process Hazards Review Chairman,
• the designer and user of the system,
• the user's supervisor,
• the Safety Engineer responsible for Process Hazards Reviews,
• the area engineering supervisor, and
• two experts in the fields of high pressure and barricades.

The committee convened in a conference room and reviewed all the material which had previously been sent to them. This included all the information previously recommended in this paper in the first part of the Process Hazards Review Agenda section as well as environmental air emissions data and gas chromatograph procedures.

The committee reviewed the material and then inspected the system. The next day, the chemist issued the minutes of the PHR on his flow reactor. These were sent to each of the participants and included the details of the time and place of the meeting; a brief review of the purpose of the PHR, namely to review the modifications; and the four recommendations that arose from the review. Minutes such as these insure that all members of the committee are in agreement, especially with the recommendations.

This review was then complete and the documents were sent to a central file.

The next example is of a Pre-Startup Review or a Process Hazards Audit held on a piece of purchased processing equipment, a Haake Rheocord Torque Rheometer and Laboratory Twin-screw Extruder. The organization responsible for this extruder has a standard checklist of eleven pages of items to be considered by the technical person or persons in charge of the equipment or process, before the review is held. Copies of the completed checklist are then sent to each committee member before the meeting for review. Generally, in a review of this type, the group will convene at the location of the equipment rather than in a conference room to go over the material provided. In this case, the committee consisted of:

- the three technical personnel responsible,
- the safety coordinators for the Department, one of whom was the Process Hazards Chairman,
- the Site area engineering maintenance supervisor,
- the Site Safety Engineer with liaison responsibility for the Department, and
- the Process Area Supervisor.

A complete discussion of all the checklist items in this Pre-Startup Review would not be possible in this paper but consideration was given to all of them. As a result, a memorandum was issued by the technical personnel to the committee outlining the recommendations made. These recommendations covered five areas and totalled 17 in all. The recommendations included the need for a splash pan on the water tank, some guards on the pelletizer and rheometer drive, correct color coding and identification of valves, and the need for an eye wash and safety shower to be installed nearby. This Pre-Startup Review took about one hour of committee time and the equipment was not operated until the recommendations were completed and the review approved.

Conclusion

While this whole review system that has been described may sound like an overwhelming and time consuming task, the Du Pont Experimental Station is convinced that Process Hazards Reviews and Audits are worthwhile and are conducting more of them every year.

Literature Cited

1. Du Pont Safety and Fire Protection Guidelines, Section 6.1, "Process Hazards Management", Feb., 1979.
2. Du Pont Safety and Fire Protection Guidelines, Section 6.4, "Process Hazards Reviews", July, 1981.
3. Du Pont Experimental Station Safety and Fire Protection Guide, Procedure 111, "Process Hazards Review (PHR)", 1/25/82.

RECEIVED November 3, 1984

Hazards Evaluation in Process Development

DANIEL P. BRANNEGAN

Central Research, Pfizer Inc., Groton, CT 06340

A hazard-evaluation program is established to obtain safety-related information for process development. This program identifies potential hazards and establishes protective safeguards by evaluating all components of the process at key stages of development. Hazard evaluation in developmental process must be sufficiently flexible to adapt to process changes. Several new critical issues regarding hazard evaluation are identified and general considerations discussed.

Process development in pharmaceutical and chemical research is itself complicated. It begins with the transfer of technology from the discovery laboratory and ends with a manufactured product, or a less-successful conclusion. Development may take several years or just a few months, during which the process undergoes frequent and significant change. Process development can be an exciting as well as a frustrating endeavor; it is seldom routine, or without hazard. Safety in process development is always an important consideration and a challenging and rewarding undertaking.

The goal of process development is to provide an efficient and safe process for manufacture. Thus, a considerable amount of very detailed safety information is obtained and designed into the final production process. At the initial stage of development (upon transfer from the discovery laboratory) there is little safety-related data available and the potential hazards are uncharacterized. Faced with a lack of safety information process development has two major responsibilities: safeguarding personnel and facilities during development and providing a safe process for production. To accomplish these objectives, an effective hazards evaluation program must be established within process development. Ideally, the hazards-evaluation program should be constructed to review each process at several key stages. This review should identify potential hazards associated with a process. These hazards should be evaluated by internal personnel with expertise in related disciplines, who establish the procedures to control these hazards. Once safeguards have been

0097–6156/85/0274–0017$06.00/0

identified, they must be effectively communicated to all concerned. Effective communication of hazard information is a critical step in the hazard-review procedure. To ensure that protective measures are well provided, these procedures should be written into all operating instructions.

Primary responsibility for hazards evaluation must reside with the line supervisor directly responsibile for the process. Although a variety of other groups may support and contribute to hazards evaluation, responsibility for the selection of tests, evaluation of hazards, and implementation of appropriate controls rests with the process supervisor.

Hazard evaluation is one component of the overall developmental process. Although hazards evaluation is important throughout development, it is both most important and most difficult to apply in the initial stages. In the very early stages of development there is little hazard information and minimal first-hand experience. During early development, the process is likely to undergo significant and frequent modifications and its most dramatic increase in scale. While the hazard-evaluation program must be *planned,* it must also be sufficiently *flexible* to evaluate the process as it changes. It is tempting, particularly early on, to attempt to obtain a considerable amount of information. Overly formalized evaluations, however, may require unnecessary tests and provide useless or misleading information. Requiring too much information early in development may provide information which is later nullified by a process change. To be most effective hazard evaluation should be performed in concert with the development of the process.

In light of these difficulties the hazard-evaluation program must establish safeguards in early stages of development. First, all *available* information related to the process should be obtained. One good source of this information is the discovery laboratories who originated the process. Hazard information observed during discovery might include: known hazardous reactions; observed exotherms; evidence of rashes; irritations, odors, etc.; and detoxification or scrubbing procedures. Handling procedures for the final product should also be described.

The second component of the initial hazard evaluation is a detailed review of the process. When a process is transferred from the discovery laboratory, a variety of hazards may appear which had earlier been well controlled by the relatively small scale of the discovery effort. Often, discovery operations cannot be translated directly into the larger scale of development. One effective means of controlling the introduction of these inherent hazards is to avoid them. The avoidance of recognized hazards is so well integrated into the transfer from discovery that its value is often overlooked. Avoiding potential hazards has become a sufficiently consistent practice, and thus, risk-reducing procedures may be identified.

For example, ethyl ether and pentane are widely used solvents in discovery. At large scale, however, they present severe potential fire hazards and developmental personnel seek to replace these substances immediately upon scale-up. Experience has shown that few processes have suffered from this action. There are a wide variety of other practices routinely employed in preliminary hazard evaluation. They include *avoiding,* where possible: evaporations to dryness; column chromatographies; reagent preparations; known reagent incompatibilities; and excessive stoichiometry.

Of course not all problems may be avoided. In the early phase of development where risks are uncharacterized, unknown hazards may be minimized through use of generic controls. Generic controls are often overprotective safeguards and may incorporate a variety of procedures as well as protective equipment. For example, reactions which contain a hazardous component that cannot be directly replaced may be run in a specially designed laboratory or in specially designed equipment.

Generic controls, though, do not provide complete protection and may paradoxically introduce new hazards. When faced with unknown hazards generic controls are widely useful, but may be overused. With safety, as with other endeavors, more is not necessarily better. Extra-precautionary measures must be utilized carefully and should be based upon information related to process hazards.

The quenching of phosphorous oxychloride may be used as an illustration of the potential hazards of improperly applied generic controls. The latent hydrolysis of phosphorous oxychloride is well known; thus, upon scale-up one is tempted to exercise extra-precautionary measures to control the hydrolysis. Some of these precautions, such as use of cold water and ice for quench, or maintaining efficient cooling with slow addition, actually can *increase* the hazards.

Phosphorus oxychloride is largely immiscible with cold water. Added reagent will "pool" in the bottom of the vessel, even with efficient stirring. Once sufficient heat has been provided by the partial hydrolysis at the water-phosphorous oxychloride interface, complete reaction of excess reagent proceeds nearly instantaneously. Often, this sinister condition may be sufficiently delayed to allow a considerable excess of phosphorous oxychloride to be added.

Any hazard evaluation must thus include an examination of *all* phases of the process. As with the example above, one area which is often ignored involves the quenching, scrubbing, and disposal of reactions. It must be remembered that these reactions suffer the same consequences of scale-up as the more "productive" components of the process.

Throughout development, but most particularly in the early stages, two factors which may introduce new hazards must be continually examined: *change and scale-up.*

Change complicates hazards evaluation by introducing new components whose hazards may be unknown. All hazards evaluation programs must be designed to accommodate the changing process, particularly in the early stages of development. Process changes not only introduce new unknowns, but may also negate previously acquired safety information. New hazards may be also introduced as a result of addressing known hazards. (Recall again the phosphorous oxychloride example.)

Scale-up, particularly initial scale-up, presents significant potential hazards and requires close evaluation. Often, the most significant risks in development are during this phase. Uneventful discovery reactions on a milligram scale can result in dramatic events on a 500gm scale (10,000X increase). Scale-ups often have their own particular hazards checklists. These checklists may include those factors most affected by the increase in scale. For example, the effect of scale-up on concentration; heat dissipation (ambient-heat loss); side reactions; and time differentials must be evaluated.

Specific Tests

While it may not be appropriate to write a schedule of tests or to require excessive testing, there are times where certain processes should always be *preceded* by specific tests. For example, certain tests may be required whenever milling/grinding or bulk drying is anticipated. These tests are designed to provide basic information on the hazards potential of products and intermediates with regard to their thermal characteristics. Such tests are termed "small-scale thermal analysis" and include the following components:

1. A burn test employing 1-3 grams of material. If appropriate, a flame test would also be performed using 50-100 grams of material.
2. Thermal decomposition test using 5 grams of sample brought to 300°C in 5 degree increments.
3. DTA/DSC (differential thermal analysis and differential scanning calorimetry).
4. Water-compatibility test.
5. Impact test utilizing 50-100 milligrams of material.

All of these are screening tests which may be performed by laboratory personnel. More sophisticated tests (such as the flame bed, JANAF, various calorimetric analyses, etc.), may be obtained through commercial testing facilities on the basis of these screens.

The fact that these screens may be performed in-house further exhibits the responsibility of the process supervisor for hazard analysis. These tests reduce large sample requirements, expense, and long analysis time which can be impediments to obtaining hazard information.

Not all testing, no matter how well planned, provides the data expected. Each test result must be carefully evaluated to obtain the maximum amount of information related to the process. At times, hazards information may be obtained from tests performed for another purpose or some other unexpected source.

As the process progresses through development, procedures designed to address specific hazards are established. As this occurs, generic controls which are overprotective give way to more precisely defined controls. The hazards-evaluation program becomes more refined as the process develops. Typically, a detailed literature search of all items of commerce used in the process is initiated. A series of physical tests are performed on isolated intermediates and products and industrial-hygiene monitoring studies begun. The hazard evaluation program may require that certain reaction sequences be run under abnormal conditions. These abnormal sequences often provide valuable information of the hazards of major equipment failures (loss of cooling; prolonged heating; agitator failure).

The results of these studies, combined with considerable experience with the process, form the basis for the technology transfer to the manufacturing facility.

Other Considerations

Hazard evaluation will be changing in the future. Right-to-know legislation, including the OSHA Hazards Communication Standard, will place great emphasis on the availability of hazard information. In this context, it is reasonable to assume that the hazards information supplied will be complete and accurate. But these laws will not quantitatively increase the amount of hazards information available and care must be taken not to *assume* that if certain hazards are not specified, they do not exist.

As science proceeds, many of our endeavors will be with new materials of increased potency and unusual physiological properties. Undoubtedly the technology necessary to produce these materials will become increasingly complex, and require more creative and informative hazards evaluation programs.

Conclusion

Hazard analysis in process development is more a philosophy than a precise program. The complex and changing nature of process development requires the scientist to be constantly on guard for the unexpected and the unknown. Such programs must be flexible as well as structured: most of all, they must become an established, well-integrated part of all development activities.

RECEIVED November 3, 1984

Risk Assessment Techniques for Experimentalists

DAVID J. VAN HORN

Research Laboratories, Rohm and Haas Company, Spring House, PA 19477

Most organizations with a written Health & Safety Policy contain statements that hazards will be identified and controlled. It is generally recognized that managers and scientists have, as part of their work, the responsibility to see that all prudent actions are taken to ensure their operations do not lead to unacceptable risks to the health and safety of the organization's employees, customers or to the environment.

There are a variety of "safety systems" available to systematically review projects to help identify hazards. However, most systems seem too laborious to be practical and/or not applicable at all for use by scientists engaged in bench research or scale-up work. This paper describes some risk assessment techniques and a mechanism for identifying hazards that are not burdensome and can readily be used by experimentalists.

WHAT PRODUCES HAZARDS

Before discussing risk assessment techniques, it is worthwhile to review what produces hazards. A recent definition of an accident by W. G. Johnson, former General Manager of the National Safety Council professional staff and author of MORT Safety Assurance Systems, provides an excellent basis for determining what produces hazards. According to Mr. Johnson, the elements involved in an accident are:

A. An unwanted transfer of energy,
B. Because of a lack of barriers and/or controls,
C. Producing injury to persons, property, or process,
D. Preceeded by sequences of planning and operational errors which:
   (1) failed to adjust to changes in physical or human factors.

0097-6156/85/0274-0023$06.00/0
© 1985 American Chemical Society

(2)   produced unsafe conditions and/or unsafe acts.
E.   Arising out of the risk in an activity,
F.   Interrupting or degrading the activity.

This may seem like a long definition but it points out hazards
may be produced by:

A.   Energy transfer
B.   Planning and operational errors or oversights
C.   Change

Routine inspections and housekeeping are valuable and important
but may not detect major hazards. The independent search out of
hazards by safety and health reviews is vital.

PRELIMINARY HAZARD ANALYSIS (PHA)

A PHA has traditionally taken the form of inventorying all the
materials and equipment to be used and deciding what are the
hazardous elements. Intuition, experience and judgement are
applied to determine what can lead to accidents and whether the
risk is acceptable or the hazard must be corrected by controls
and/or contingency plans. However, a more thorough review can
be achieved with the following PHA procedure.
To begin research, the experimentalist must thoroughly think
through the entire process, step by step, anticipate what might
inadvertently go wrong, how to prepare for same, or what to do
if the worst happening occurs. Planning for safety, health, and
environmental considerations is an integral part of research and
shall:
  1. Identify potential hazards
  2. Take measures to minimize risks to an acceptable level
  3. Make preparations to handle any mishaps that may occur
  4. Insure instructions to all involved are understood and
     followed.
A good PHA will speed development by preventing serious accidents and providing answers to questions that will be raised
later. The descending priority in developing a safe process is
one that is:

  1. Intrinsically safe (inherently safe regardless of
     external circumstances)
  2. Safe by use of engineering/design controls (containment,
     ventilation, guards, etc.)
  3. Safe by use of administrative controls (SOP, qualified
     personnel, maintenance, etc.)
  4. Safe by use of personal controls (respirators, eye/face
     protection, gloves, etc.)

A checklist for experimentalists and managers to use for a PHA
is provided in Appendix 1. Remember, many accidents occur as a
result of a combination of unusual, improbable circumstances.
The PHA checklist includes reminders to use in-house services
like the Library literature search, Analytical, Computer Appli-

cations (ASTM, CHETAH Program) and Health and Safety Staff
department resources.

The PHA may sound like a lot of review to identify hazards
but many still slip by or are created for a variety of reasons.
One reason is not being sure all the important questions have
been asked.

Following are some simple additional techniques to help
gather information and systematically identify risks for consideration.

SIMPLE RISK ASSESSMENT TECHNIQUES

1. Incident Recall

This information gathering technique is also known as "critical
incident technique" and is a means of collecting both poor
and good experience from operationally experienced personnel.
It requires asking people to share difficulties, errors, near
misses, accidents, successes etc. they remember in past similar
operations and conditions.

This method has generated a greater quantity of relevant
and useful information than any other monitoring technique. It
identifies more seemingly minor errors or deficiencies and near
misses. This is to be expected as the familiar Heinrich Tri-
angle shows there are numerous near misses for every accident.

2. Failure Modes and Effects

This procedure is sometimes called the "What If" technique and
helps identify items affecting process reliability by considering each potential source of unwanted energy flow and identifying the failure modes by which release can occur and the resulting effects on the system. This is usually done by making a
line diagram of the process, breaking it down into subsystems if
necessary, and studying all modes of operation. A procedure to
follow is to consider:
  A. Failure or error mode of each component.
      1. Instrument, equipment failures.
      2. Supply, delivery failures.
      3. Human failures.
      4. Abnormal operations, change from routine.
      5. Emergency and major environmental events.
  B. Effect of failure on other components and the whole system.
  C. Whether hazards identified are serious and probable.
  D. Methods of failure detection.
  E. Provision for contingencies.

Unless the process under consideration is relatively simple, this
type analysis can be complicated considering all the energy
sources and interactions that can occur. This can lead to double-
failure and Fault Tree Analysis.

Change Analysis

Change may be the mother of progress but it is really the
mother of twins; progress and trouble. The consequences of
change should always be considered carefully. Many accidents
are caused because the results of changes are not anticipated.
The procedure for change analysis is:
1. Consider the usual operation (accident free situation).
2. Consider the change(s) to be made.
3. Compare the new situation with the accident free reference.
4. Analyze the differences for effect on producing an accident.
    This must be done with careful attention to obscure and
    indirect relationships.
5. Integrate information developed relative to causative
    factors into the accident prevention process.
6. Determine if any hazardous material testing is needed in
    order to define safe operating parameters.

Job Safety Analysis

JSA is a technique for the review of a job to uncover inherent
or potential problems. It also is an excellent tool to help in
developing a good SOP and for orienting the new employee and in
training programs. It involves both the supervisor and the
employee(s) doing the job so both contribute and learn.
The job selected for analysis should be recognized as having
the potential for serious accidents and be relatively standardized. The steps are:
1. Select the job
2. Break the job down into its sequential steps
3. Identify the potential hazards of each step (consider
    energy sources, watch work practices)
4. Establish controls for the identified hazards
5. Evaluate the controls (make sure they are implemented into
    operational systems).

MORE SOPHISTICATED RISK ASSESSMENT TECHNIQUES

1. Safety, Health, Environmental Review (SHE)

A SHE review is required whenever the experimentalist plans
work involving unfamiliar chemistry, toxicity or equipment, or
plans the scale-up of a process, or plans for toll or contract
research outside the company.
     A SHE review may involve only the experimentalist and
department supervision or be expanded as needed to include
other laboratory department and/or staff personnel that can
contribute. The same questions can be asked as in the PHA
checklist for simple department reviews.
     For more involved reviews or for the scale-up of a process,
a more formal procedure should be followed and more detailed
checklists are available. A suggested SHE review format and
definition of scale-up is provided in Appendix 2.

2. Hazard and Operability Study (HAZOP)

The basic concept of HAZOP is to:
    a. Fully describe the process and break it down into logical parts which can be considered separately.
    b. Systematically question every part to discover how deviation from what is intended can occur.
    c. Decide whether any deviation can create hazards.

The purpose of a HAZOP review is to identify hazards before an incident - not necessarily to solve how to eliminate or minimize the potential hazard. Some solutions are obvious and can be handled immediately. However, many solutions may be complex and should be assigned to others to solve outside of the HAZOP review process.

HAZOP can be very flexible. It is valuable in the design stage and in assessing the hazard potential of operational failures of individual items and the consequential effects on the whole.

Even if you never participate in a formal HAZOP review, the principles are extremely valuable in everyday operations as it broadens thinking to include more potential problems and to handle them in a systematic way.

    Some necessary definitions are:
1.   Intention - define what is expected
2.   Deviation - systematically question how deviations can occur
3.   Causes - are the reasons why deviations might occur
4.   Consequences - are the results of deviations should they occur
5.   Hazards - are the consequences which can cause injury, illness or loss.

    The following guide words are used for each part identified in the process and applied to each intention to assist in discovering possible deviations from the intention:

3. Fault Tree

Fault tree analysis is many times requested because of complex hazards identified by a HAZOP review. Engineering has some people trained in analytical tree development and analysis and they are responsible for conducting these type studies.

    Briefly, a fault tree is designed by setting down the undesired event at the top and determining all the specific events which can bring about the failure. These events are tiered below. Then it is possible to calculate or estimate the failure probability for each event. This is normally done for the life cycle of the operation.

    This quantifies the events and gives an idea what the greatest risks are and where to make changes to give the greatest additional safety for the money available.

GUIDE WORDS
_____

GUIDE WORDS	MEANING	EXAMPLES
NO OR NOT	NO PART OF THE INTENTION IS ACHIEVED BUT NOTHING ELSE HAPPENS.	NO FLOW, NO AGITATION, NO REACTION
MORE LESS	QUANTITATIVE INCREASE OR DECREASES TO THE INTENDED ACTIVITY.	MORE FLOW, HIGHER PRESSURE, LOWER TEMPERATURE, LESS TIME
AS WELL AS	ALL OF THE INTENTION IS ACHIEVED BUT SOME ADDITIONAL ACTIVITY OCCURS.	ADDITIONAL COMPONENT, CONTAMINANT, EXTRA PHASE
PART OF	ONLY PART OF THE INTENTION IS ACHIEVED, PART IS NOT.	COMPONENT OMITTED, PART OF MULTIPLE DESTINATIONS OMITTED.
REVERSE	THE OPPOSITE OF THE INTENTION OCCURS.	REVERSE FLOW, REVERSE ORDER OF ADDITION
OTHER THAN	NO PART OF THE INTENTION IS ACHIEVED. SOMETHING DIFFERENT HAPPENS.	WRONG COMPONENT, STARTUP, SHUTDOWN, UTILITY FAILURE.

4. Management Oversight and Risk Tree (MORT)

MORT is an even more sophisticated program for managing safety systematically by use of logic trees. It is currently used for major government projects in DOE, NRC, etc. for project review and startup and in the investigation of serious accidents. It is complex because it not only includes the technical aspects of a fault tree analysis (hardware) but includes logic trees for the deductive analysis of managerial functions, human-behavioral factors, and environmental considerations. Obviously, only large complex projects with numerous inherent hazardous operations that could result in serious consequences, if there were failures, can afford to be analyzed by MORT.

SUMMARY

Experimentalists can consider the safety, health, and environmental consequences of their planned work in a systematic way by:
1. understanding their responsibility to identify hazards,
2. knowing what produces hazards,
3. using a Preliminary Hazard Analysis checklist,

4. utilizing risk assessment techniques, and
5. applying the proper level of analyses based on the
    complexity of the operation and possible consequences of
    failure.

APPENDIX 1
PRELIMINARY HAZARD ANALYSIS CHECKLIST

1. Conduct literature search but remember accidents and
    unusual results are not always reported.
2. List possible reactions and side reactions. Can less
    hazardous chemicals be substituted to achieve desired
    results?
3. Obtain MSD Sheets or if not available, contact the
    Industrial Hygienist. Review the characteristics of
    all reactants, intermediates and products in terms of
    flammability, toxicity and reactivity hazards. Where
    information is not available, treat the materials as
    hazardous.
4. What is the flash point, flammability range, auto-igni-
    tion point, vapor pressure and density?
5. What is the threshold level and type hazard (inhalation,
    ingestion, skin contact)? What protective measures are
    required?
6. What is the recommended first aid in case of accidental
    exposure?
7. Will work require radiation or noise control measures,
    monitoring for biological or chemical air contaminants,
    or medical surveillance?
8. How much material/energy is involved and how violent
    may the reaction be? Consider use of the CHETAH
    system. Determine quantity and rate of evolution of
    heat and gases. Consider use of ARC.
9. Does it decompose and if so, how rapidly, and to what?
10. Is it impact sensitive?
11 What is its stability on storage to cold, heat, light,
    water, metals, etc.?
12. What are effects of catalysts, inhibitors, or contami-
    nants (like iron) on the reactions?
13. Will water or air affect the reaction?
14. Can mischarge or wrong addition order affect the reac-
    tion?
15. Are incompatible chemicals involved or likely to be
    generated?
16. Will work require special precautions to prevent odor
    problems, air pollution, or sewer contamination?
17. Can wastes be safely handled and arrangements for
    disposal completed?
18. Does equipment fit safely into area allocated? Need
    isolation, shielding, pressure relief, ventilation,
    redundant controls, automatic shutdown, etc.?
19. Can all parts of the system be vented before breaking
    any lines?

20. What would happen and what should be done if:
    - Electric power fails?
    - Cooling or heating system fails?
    - Automatic controls or equipment air fails?
    - Ventilation fails?
    - Pressure gets out of hand?
    - Water or air leaks into system?
    - Material or reaction container falls and
      breaks or spills contents?
21. Have personnel who may be involved been notified of any special
    hazards or precautions: neighbors, services, medical, emergency
    response personnel, etc.

APPENDIX 2
SAFETY, HEALTH, ENVIRONMENTAL REVIEW PROCEDURE

A SHE Review will be held when requested as a result of a
department PHA or agreed necessary by others such as department
supervision, Research Engineering or Health and Safety, Corporate Engineering or Health and Safety, etc. In addition, there
are procedures established by Pilot Plants, Corporate Depart-
ments, and in the manufacturing sites that require formal
Safety Reviews for new facilities and before research processes
are scaled-up in the Semi-Works, Preps Lab, Pilot Plant or go
to plant production.
Scale-up is a term used when going from small glassware (100 ml
or smaller) to larger glassware (1-5 liters) or to even larger
glassware (12-22 liters). It is most frequently used when
going from glassware (1-5 liters) to equipment in formulation
laboratories or pilot plants (generally 5-10 gallons or more).
A SHE review can be held, if requested by the Scientist or
Manager, if the work stays in glassware and in the department.
A SHE review is required if the scale-up work is transferred to
another department or goes from glassware to gallon sized pilot
plant equipment.

1. Responsibilities for SHE Reviews

a. The line manager responsible for the project (Department
   Manager, Senior Research Associate or Section Manager) has
   the responsibility to determine when a SHE Review should be
   held, decide the Committee membership and call the Meeting.
b. The senior line manager on the Committee should act as
   Chairman.
c. A S/H Monitor or person appointed by the Chairman will keep
   and issue minutes outlining action points and persons
   responsible to incorporate Committee recommendations.
d. The Chairman is responsible for decisions when agreement is
   not reached in committee.

2. Membership - The composition of the SHE Review Committee
   can vary considerably depending on the project. A suggested membership for a large formal review is:

a.   Appropriate Research Director - invite to all meetings.
b.   Appropriate Department Manager/Senior Research Associate.
c.   Section Manager and Chemist(s) or Engineer(s) associated
     with the work.
d.   Department Safety/Health Monitor.
e.   Member of Research Health and Safety Department staff.
f.   Any other personnel that can contribute to the meeting such
     as those in Research Engineering or Toxicology Departments
     with specialized knowledge, or Corporate Engineering or
     Health and Safety, Plant Safety, Engineering, or Environ-
     mental Control, etc.

Literature Cited

1.   MORT Safety Assurance System, William G. Johnson, Marcel
     Dekker Inc., New York & Basel, 1980, pgs.23, 58.

RECEIVED December 4, 1984

Hazard and Operability Study

A Flexible Technique for Process System Safety and Reliability Analysis

A. SHAFAGHI and S. B. GIBSON

Columbus Division, Battelle Memorial Institute, Columbus, OH 43201

Safety and reliability of chemical process plants are such important issues, they deserve the best techniques to prevent problems occurring. To minimize risks resulted from operating problems and hazardous events, process system safety and reliability analysis is often employed. This is a rigorous approach undertaken to improve system reliability and safety. The approach consists of three main tasks; hazard identification, risk estimation, and risk control. The first task is crucial in process system safety analysis, because the effectiveness of the other two tasks depends on it. The technique of hazard and operability (HAZOP) study is a systematic approach to identifying most potential hazards and operating problems. The technique in contrast to the traditional methods is simple, creative, and flexible.

During the 1960's, the chemical industry developed rapidly, and to achieve the benefits of scale, chemical plants became larger and more sophisticated. As communications and greater public awareness heightened the effect of incidents in the industry during the 1970's, great strides had to be made to improve the tools available to increase process safety and reliability. Its threefold purpose is described below.

The first task is to identify potential hazards and operability problems. In general, there exist two types of hazard: inherent, due to the nature of raw materials used; and subtle, due to omissions and errors made in design.

After the hazards have been identified, the second task is to determine the risks associated with them. Risk is defined as the likelihood of the hazardous event and the severity of the accident. Fault trees are often used to quantify the likelihood of hazardous events. The severity is usually defined as the degree, sometimes in terms of likelihood, of exposure to accidents.

0097–6156/85/0274–0033$06.00/0
© 1985 American Chemical Society

The ultimate goal in process system safety and risk analysis is to control the risks. This final task is carried out by comparing the risks calculated with risk criteria specified by an authority. The criteria can be subjectively determined, based on the past experience or the existing background risks. However, many companies (1,2) have established numerical targets for risks. Given calculated risks beyond the specified limit, decisions will be made to improve the design or operation and maintenance procedures to reduce the risks.

The first task, hazard identification, is crucial in process system safety analysis, because the effectiveness of the other two tasks depends on it. The traditional methods for identifying hazards during the 1960's (including 'process reviews', 'codes of practice', 'checklists', and 'safety audit') were no longer considered adequate in the 1970's. There was a need for a technique which could anticipate hazardous problems, particularly in areas of novelty and new technology where past experience was limited.

The technique of hazard and operability (HAZOP) study was developed to fill this need. HAZOP study has the following noteworthy features:

- It is based on brainstorming
- It takes a multidisciplinary team approach
- It is structured by using guide words
- It is cost effective.

A brief comparison between the traditional methods (such as the "what if" method) and HAZOP study is given in Table I.

Table I. A Comparison Between Traditional Methods and HAZOP Study

Traditional Methods	HAZOP Study
Experience-based	Creative
Procedural	Systematic
Collective	Collective with constructive interactions

Conducting a HAZOP Study

The approach to identifying hazards and operability problems by this technique is to search for deviations from design intents. The first step is to plan the HAZOP study to ensure that there is sufficient time, expertise, and information available. Next, a multidisciplinary team, led by an expert in the technique, is set up. Then, by means of a fixed set of abstract words called guide words (such as 'more'), the team leader examines each process

variable or parameter of interest (such as 'flow') and conveys the
deviations (such as 'more flow') to the team members. The objec-
tive is to stimulate the members into creative thinking about the
consequences and causes of the deviations. Finally, the team
agrees on possible causes, consequences, and solutions of the
problems posed by the deviations and recommends any further
actions to be taken.

The steps in conducting a HAZOP study are described in
greater detail below.

Planning the Study. Planning HAZOPs is an important task, and
complicated by the following factors:

- HAZOPs are time consuming and should be done during the
  design phase--normally a busy time for the participants
- HAZOPs should be planned well in advance, because they are
  a big commitment and a heavy load for each participant
- The piping and instrument (P&I) diagrams must be in a
  sufficiently finalized form.

Experience has shown that, particularly on large projects,
conducting a preliminary hazard analysis earlier in the project is
a considerable help in integrating the HAZOPs into the project
plan (3). The preliminary hazard analysis will seek out the
major, obvious hazard and reliability problems.

Assembling the Team. For a chemical process at the design stage,
the team members would include the process engineer, the research
chemist, and the mechanical design engineer. On a highly auto-
mated process, the instrument and control engineer might be
included, and advisors on explosion hazards, toxicology, and materials of construction may be present when required. An effective
team size is 5-7 people, and it is important that all members
contribute.

A team leader is necessary to ensure that the methodology is
properly applied and to keep the meeting under control. This
latter task should by no means be underestimated.

Applying the Guide Words. HAZOP study is a free-wheeling technique, but it is structured by means of seven abstract words,
called guide words. The guide words used in HAZOP studies,
together with their meanings, are given in Table II.

Table II. The Guide Words (4)

Guide Words	Meanings
No, None	Negation of the design intent
More	Quantitative increase
Less	Quantitative decrease
As well as	Qualitative increase
Part of	Qualitative decrease
Reverse	Logical opposite of the intent
Other than	Complete substitution

These guide words are applied to process variables and parameters of interest for the system under study. On continuous
chemical processes, process variables include temperature,
pressure, flow, and concentration. For example, a set of specialized guide words (deviations) for these variables are given below:

Guide Word	Variable	Deviation
No	Flow	No flow
Less	Temperature	Low temperature
More	Pressure	High pressure
Part of	Concentration	Low concentration

For batch processes, level, reactivity, and time might be
additional parameters considered.

For electrical systems, voltage, current, phase, and frequency are among the parameters to be considered.

Finding the Results. To conduct the study (5), the HAZOP team
examines P&I diagrams. Each diagram is divided into discrete
homogeneous sections, usually pipes, and each section is considered in turn. Each of the guide words is applied to the
section being studied to stimulate the team into imagining what
could go wrong with any part of the system given the suggested
deviation at that section.

One of three outcomes is possible for each guide word
application:

(1) No hazard or problem exists

(2) A hazard or problem exists. In this case, a suitable
     record is made to that effect, and the requisite
     solution will have to be found outside the meetings

(3)   The team does not have sufficient information to determine whether a problem exists. In this case, a record
is made to that effect, and again the necessary information will have to be found outside the meeting.

When each guide word has been applied to each section of each
diagram, the HAZOP is complete; but of course all the questions
recorded still have to be resolved.

The method is simplified and shown graphically in Figure 1.
Note that the operation in the decision box in fact includes
several iterations, each for one specialized guide word.

Benefits

The major contribution HAZOPs make to a project's effectiveness is
that they identify potential problems at the design stage rather
than when they become incidents. This provides a number of
benefits:

- Potential problems are resolved relatively easily, and
  most subtle hazards are identified at the design stage
- Potential problems can be resolved rationally, whereas an
  incident usually creates an overreaction and expensive,
  ultra-conservative solutions
- Engineering change orders during construction and com-
  missioning are drastically reduced
- Startup is more timely
- Plants reach design rates quicker.

As an example of the cost effectiveness of the technique, an
assessment of the value of HAZOPs on a $38 million project was
done. Table III shows the cost effectiveness of the HAZOP study
technique for a new project clearly.

Flexibility

As we have seen, the methodology is basically very simple, and
because the guide words are general, the HAZOP can be applied to
any type of system.

Some of the systems studied include:

- Continuous chemicals and petrochemicals processes
- Batch organics, specialty chemicals and pharmaceutical
  processes
- Pilot plants
- Bench research processes
- Electrical interlock systems
- Computer installations
- Drainage systems
- Molecular genetics research laboratory.

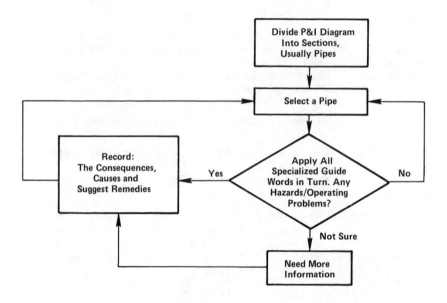

Figure 1.   The Method of HAZOP Study.

Table III.   Plant HAZOP Study

	$000's	%
Total plant capital cost	38,000	100
Cost of HAZOP study	60	0.2
Cost of modifications to problems revealed by HAZOP study	647	1.7
Cost of corrections if study not done	1,487	3.9
Capital savings	780	2.1
Other savings on operating costs where problems could not have been corrected practicably after the design stage	262/yr	0.7

HAZOPs can be applied equally well to new designs or existing situations, although in the latter case solutions to the potential problems identified by HAZOP are usually more difficult and costly to implement.

Areas where HAZOPs could be applied effectively include:

- Manufacturing processes
- Maintenance activities
- Product liability.

Conclusion

The hazard and operability study is a simple and effective technique for identifying potential problems in a wide variety of systems and activities.

Its advantages over more traditional approaches are that it is more creative and rigorous and is particularly well suited to new technologies and areas where some novelty exists.

Although easy to apply, conducting HAZOPs requires training and experience to select the best approach, plan the analysis, get the best out of the team, and record the results.

Literature Cited

1. Helmers, E. N.; Schaller, L. C. Plant/Operations Progress, Vol. 1, No. 3, July, 1982, p. 190.
2. Gibson, S. B. Chemical Engineering Progress, Vol. 76, No. 11, November, 1980.
3. Gibson, S. B. I. Chem E. Symposium Series, No. 47, 1976.
4. Chemical Industries Association, "A Guide to Hazard and Operability Studies", Publications Department, Alembic House, 93 Albert Embankment, London, SE1 7Tu, England, 1977.
5. Lawley, H. G., Chemical Engineering Progress, Vol. 70, No. 4, April, 1974, p. 45.

RECEIVED November 3, 1984

Hazard Avoidance in the Processing of Pharmaceuticals

JOHN R. HANDLEY

Sterling Organics Division, Sterling Drug Inc., Rensselaer, NY 12144

The batchwise processing of pharmaceuticals always
presents the possibility of serious physical hazard.
These hazards can usually be avoided with the help of
a comprehensive hazard evaluation program such as the
one recently installed in our facility. This program
involves the physical testing of a process as well as
a detailed examination by a Hazard Review Team (consisting of an R&D chemist, safety officer, pilot
plant director, environmental co-ordinator, process
engineer and hazard evaluation chemist). Physical
testing usually entails an examination of heats of
reaction, minimum decomposition temperatures, relative
thermal stabilities and flammability characteristics
associated with a particular process. Some specialized testing (ARC, DSC, etc.) may also be carried out
when indicated. We have found that this program is
already affording benefits of smoother and safer
plant operations.

The Sterling Organics Rensselaer Plant is a medium-sized chemical
facility producing pharmaceuticals and their intermediates as well as
some specialty organic chemicals. Production is generally on a batch
basis in medium to large reactors.
    The Sterling Organics UK plant is similar and is where, about
seven years ago, the concept of in-house hazard evaluation was
conceived.
    The beginnings of our hazard evaluation program were fairly low
key. For years we had been running a Schiemann reaction which
involved the thermolysis of an aromatic diazonium fluorborate salt
(Figure 1).

0097-6156/85/0274-0041$06.00/0
© 1985 American Chemical Society

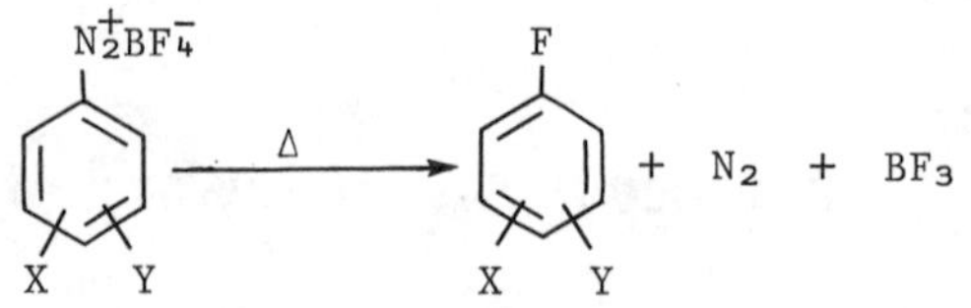

Figure 1.   Schiemann Reaction

In this reaction the starting salt was heated to about 90° at which
point it began a slow, clean liberation of gas. One particular
batch of starting material, however, decomposed vigorously at an
unexpectedly low temperature with rapid evolution of gas and heat.
Although no serious damage took place, it was apparent this familiar
reaction was indeed not safe even though it had run for years in the
lab and plant without incident.

Many reactions cannot be deemed safe for full scale operation
in the plant even though no problems were evident in the lab.

The primary reasons for this difference in reactive characteristics are the scaling effects associated with much larger batch
size. Simple mass transfer effects make cooling a larger batch a
much slower and more difficult process. Hot spots on the wall of
the reactor vessel or during reagent addition in a reactor may also
be potential sources of difficulties. Contamination in production—
iron, for example—is not a significant problem in a laboratory
environment, but must be actively excluded in the plant. Finally,
even a small kettle has a large degree of adiabaticity relative to
a 5-liter flask. A reaction that generates even a small amount of
heat can thermally self-accelerate if an adequate heat-removal
system is not available.

The starting material which unexpectedly decomposed was
examined in the lab. Samples of this particular batch of material,
as well as batches of material that had reacted normally, were
subjected to heating in an oil bath at 2-3°C per minute and the
initial decomposition temperatures observed. The temperature
difference between the sample and oil bath was plotted against the
oil bath temperature. With this crude equipment, it was shown that
the vigor of the exotherm was directly related to the quality of the
fluoroborate. Thus, a pure sample of the fluoroborate (Figure 2)
decomposed at 95°C with smooth evolution of gas and no foaming.
With an impure sample the batch decomposed violently at 84°C (Figure
3). As a result of this work, strict controls were placed on the
washing and drying of the fluoroborate salt. In addition, each
batch was subjected to this simple test to determine its response to
heating before the plant decomposition was performed. These safeguards allowed the safe handling of this reaction on the plant scale.

Initial Hazard Evaluation Objectives

The success of this original, crude work inspired us to develop more
refined testing procedures which utilized existing or relatively
inexpensive equipment. Eventually, a formalized hazard testing

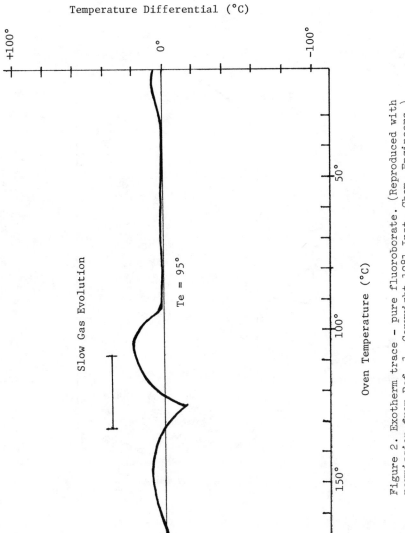

Figure 2. Exotherm trace – pure fluoroborate. (Reproduced with permission from Ref. 1. Copyright 1981 Inst. Chem. Engineers.)

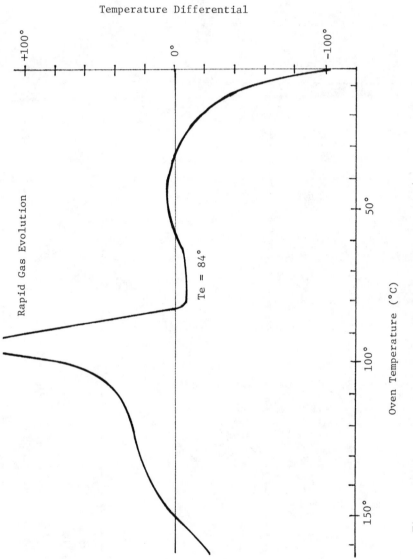

Figure 3. Exotherm trace – poor quality fluoroborate. (Reproduced with permission from Ref. 1. Copyright 1981 Inst. Chem. Engineers.)

program was developed with the following objectives:

1. Establish a laboratory to examine the hazard potential of a particular reaction or compound, initially using simple tests and equipment.
2. Provide data on the potential for uncontrolled exotherms to R & D and Production groups.
3. Examine all new products entering both Production and the Pilot Plant for possible unexpected hazards.
4. Finally, and probably most importantly, educate and train chemists, engineers and managers on the role of the hazards lab in achieving safer in-plant operating conditions.

Two priority schedules were developed to determine in what order compounds and processes should be tested. The first priority listing covered the testing which was needed in a short time, that is, those processes which could pose a real, immediate threat to plant personnel. These included:

1. All plant processes where concern existed regarding the reaction's controllability or reactions which had, in the past, shown any evidence of variable exothermic behavior.
2. Residues from distillations of products or intermediates which were subjected to temperature in excess of 250°C.
3. Raw materials, intermediates or products which contained functional groups known from experience to have potential for instability.
4. All plant processes involving nitrations or strongly oxidizing conditions and processes with high rates of gas evolution.

After completion of this stage of our testing program, long-term priorities were addressed. These are:

1. Examine all processes entering the Pilot Plant for the first time.
2. Examine all processes currently in operation in the plant.

Hazard Review Team Organization

Recently, a formalized hazard assessment procedure has been initiated for all processes to be run in the Pilot Plant. To date, the testing methodology has been found to be quite effective at identifying potential hazards. For the remainder of this paper, I will elaborate on this procedure, detail its step-by-step format as well as the responsibilities of the individuals involved.

The hazard evaluation program requires the expertise of a number of different disciplines as well as the coordination and reconcil-iation of the project schedule with factors such as equipment suitability, personnel, training and effluent considerations. Obviously, to take into account all of the difficulties associated with starting and running an unfamiliar process in addition to examining the potential hazards of the process is a complicated task. The format described here will work for most manufacturing operations.

The responsibility for hazard assessment for new product introduction lies with the Hazard Review Team. The Hazard Review Team for a particular project generally consists of the following people:

1. The Pilot Plant Director who coordinates hazard review scheduling, prepares a hazard review summary and keeps comprehensive records of all information pertinent to a process.

2. The R & D Chemist most intimately involved in the project. He
   provides a basic reactive sequence summary, offers his opinion
   as to the relative hazard potential of specific operations and
   chemical species, and prepares formalized process directions.
3. The Hazard Evaluation Chemist quantifies the actual potential
   hazards involved in an operation - heats of reaction, gas
   evolution rates, minimum decomposition temperatures, etc.
4. The Process Engineer is involved when an actual potential hazard
   is identified. He suggests methods for minimizing or eliminating
   such a hazard.
5. The Environmental Coordinator decides the ultimate fate of solid
   and liquid waste products from the process.
6. The Plant Safety Officer identifies the toxic hazards of chemical
   species involved in a particular process. He also recommends
   what protective equipment may be required to preclude personal
   injury.

Hazard Review Team—Process Examination

Figure 4 is a flow chart outlining the usual progression of information and responsibility for a typical hazard review. Briefly, the
hazard review process is as follows:
    When a decision is made to pilot a process, the Pilot Plant
Director schedules the review sequence and notifies the R & D Chemist
who provides a process summary to individual members of the Hazard
Review Team. The Safety Officer will then examine the process for
the presence of possible toxic or irritating compounds while the
Environmental Coordinator decides the ultimate fate of any process
waste streams. The Hazard Evaluation Chemist identifies any thermal
or reactive hazards associated with the process. If a hazard is
identified, the Process Engineer determines the suitability of a
particular reactor grouping to handle the specific conditions
involved.
    Once every member of the Review Team has had a chance to
examine the process, the entire Team meets to decide upon the safest
operating procedures. From the notes of this meeting, a Hazard
Review Summary is prepared. The R & D Chemist uses this Summary to
prepare a finalized set of manufacturing directions. At this point
the process is considered ready for piloting.
    An important controlling feature of the hazard review procedure
is the Hazard Assessment Form (Figure 5). This Form is simply a
checklist by which individual responsibilities and suggested
completion dates are assigned for the following items:
1. Hazard and operability assessment
2. Process hazard evaluation
3. Implementation of reactor modifications
4. Effluent discharge considerations
5. Safety and toxicity screening
6. Special material handling considerations
7. Hazard review summary
8. Approved Batch Record preparation.
    An individual signs the Hazard Assessment Form only when he is
satisfied that a process can be run safely in the plant.

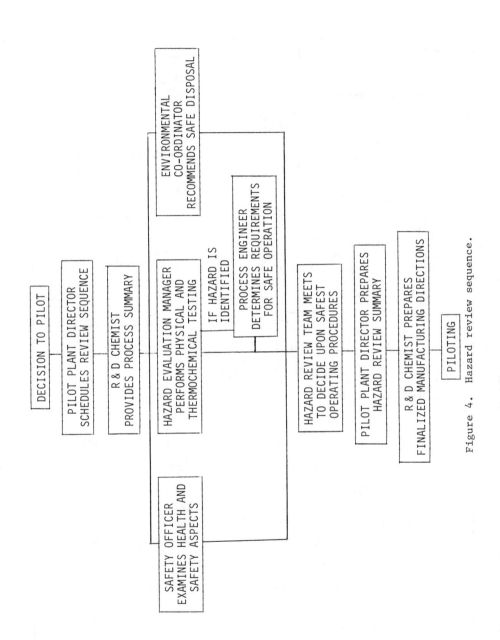

Figure 4. Hazard review sequence.

Process_____   Stage_____

1. Hazard & Operability Assesment
   Questionnaire completed - Copy of this   R&D Chemist
   form and questionaire to Hazard Review
   Team and W. H. Thielking

                                                     _____
                                                     Date_____
   Due_____

2. Process Hazard Evaluation Completed
   Potential hazard(s) identified   Yes/No

   Due_____   Hazard Evaluation Manager
   Report cc: Hazard Review Team
                                                     _____

                                                     Date_____

3. A. Plant modifications designed &   Process Engineer
      implemented (if necessary)

   B. Effluent discharge examined &
   appropriate disposal methods   _____
   determined   Date_____

   Due_____   Environmental Coordinator

                                                     _____

                                                     Date_____

4. A. Safety/Toxicity screening completed

   B. Special considerations required for   Yes/No
   safe handling of materials?

   C. Corporate notification   Safety Officer

                                                     _____
   Due_____
                                                     Date_____

5. Hazard Review Summary prepared   Pilot Plant Director

                                                     _____
   Due_____
                                                     Date_____

6. Approved Batch Record prepared   R&D Chemist

                                                     _____
   Due_____
                                                     Date_____

After Step 6, return to Pilot Plant Director

Figure 5.   Hazard Assessment Form.

Without this Form it would be quite difficult to maintain any kind of schedule. Even so, it can only be regarded as a guideline due to the vastly differing amount and type of work required for each project that comes under scrutiny.

The Hazard Assessment Form is issued by the Pilot Plant Director at the time a process is brought to him for placement on the Pilot Plant schedule. The Pilot Plant Director is responsible for assigning suggested completion dates and scheduling the overall hazard review process. He acts as the coordinator for the overall review sequence and personally keeps on file all correspondence and information pertinent to a particular process. The Form is issued to the R & D Chemist most familiar with the chemistry of the process involved.

The R & D Chemist then completes a hazard and operability assessment questionnaire which we have developed to assist him in summarizing the salient details of a particular process. This questionnaire consists of five sections in which the Chemist provides:

1. A step-by-step process summary detailing the individual operations required for a particular chemical transformation. The Chemist will also provide any pertinent chemical structures which are primarily of use to the Hazard Evaluation Chemist.

2. A tabulation of the relative potential hazards associated with particular variations from the correct process conditions, e.g., charging a reagent too fast, loss of agitation, too much heat, etc. We have found that this section of the questionnaire is not quite as useful as one would first suspect when straightforward operations are involved. When long, complex, multi-step procedures are examined in this fashion, however, the most critical points of interest are immediately highlighted.

3. A summary of any liquid or solid waste (filter cakes, washes, mother liquors or distillates) which will be produced in the process. The composition of each particular waste stream is given as well as an indication of any associated disposal problems of which the chemist may be aware.

4. Recommended safe shutdown procedures in the event of a reaction problem or a building emergency requiring evacuation. This section is filled out only if special shutdown procedures are required that are not covered in our standardized format.

5. A summary of the potentially serious process deviations which could result in problems at each step. Although this may seem redundant at first, it has been useful for simply putting the whole process in perspective. The chemist can also put down any general impressions about the process which may not have been brought out in the other sections.

After this questionnaire is completed, copies of it, as well as the original Hazard Assessment Form, are sent to the individual members of the Hazard Review Team. It is at this time that the other members of the Team are made aware of the project and their respective responsibilities concerning it. The Hazard Evaluation Chemist is responsible for clarifying the degree of hazard involved with a particular operation or material. He carries out a series of tests which are dictated by the particular chemistry involved.

Physical Testing

Our testing program (outlined in Figure 6) has recently been stream-
lined through the purchase of an Accelerating Rate Calorimeter (ARC).
This instrument has proven to be quite useful for determining
stability criteria in those areas where our previous testing methodology yielded somewhat imprecise answers.

When a process is accepted for examination, an inspection of
the structures involved is performed. Any compounds with notor-
iously reactive substituents (e.g., nitro compounds) are, of course,
subjected to ARC testing. Next, an oxygen balance is calculated to
determine if enough oxygen is contained in the molecular structure
to support combustion in a confined environment. If this is the
case, the material in question is subjected to ARC testing. Finally,
if no obvious hazard is detected, a train firing test is performed.
In this test, a thin strip of material is ignited at one end and the
rate of burning is observed. A sample that is highly flammable is
also subjected to ARC testing.

If no obvious hazard is observed at this point, a crude
stability (DTA, Figure 7) test is carried out. This test is a more
refined version of the initial tests run in an oil bath. This test
will provide an indication of the decomposition temperature of a
reaction mixture or compound. Generally we feel that if this
decomposition temperature is more than 100°C higher than the maximum
temperature to which the material will be subjected in the plant, the
process can be run with minimum danger of thermal decomposition. If
this temperature difference is less than 100°C, the material is
subjected to ARC testing.

In many cases it is necessary to determine heats of reaction.
Usually these are determined using a simple calorimeter (Figure 8).
This data is quite useful in determining cooling requirements.
Often, a reaction simulation is carried out in which a Dewar flask
is charged with a volume of reactant proportional to what would be
in a full size reactor. In this way, it can be determined if
foaming or overheating will be a problem if there is an overly fast
addition of reagent.

When testing on a particular process is completed, the Hazard
Evaluation Chemist prepares a report identifying any reactive
problems which were observed. He also reports the actual values of
any physical parameters (heat of reaction, decomposition temperature)
which were determined and makes recommendations as to what areas (if
any) should be addressed by the Process Engineer.

Engineering and Health Considerations

The Process Engineer then makes recommendations concerning the
ability of a certain reactor group to handle the reaction. He
designs and implements any changes necessary to run the process
safely in the plant. These changes usually involve simple re-piping
of a kettle or installing specialized equipment (e.g., flow meters)
to provide better control over reaction conditions. In extreme
circumstances he may recommend the installation of an "idiot-proof"
system (such as an orifice plate in a line for limiting addition
rates).

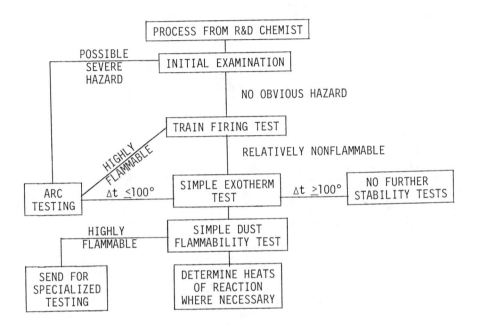

Figure 6. Hazard testing sequence.

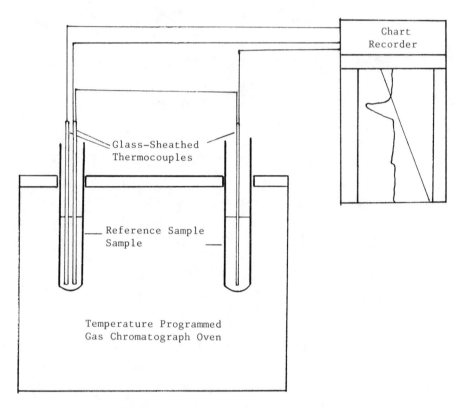

Figure 7.   Simple exotherm apparatus.

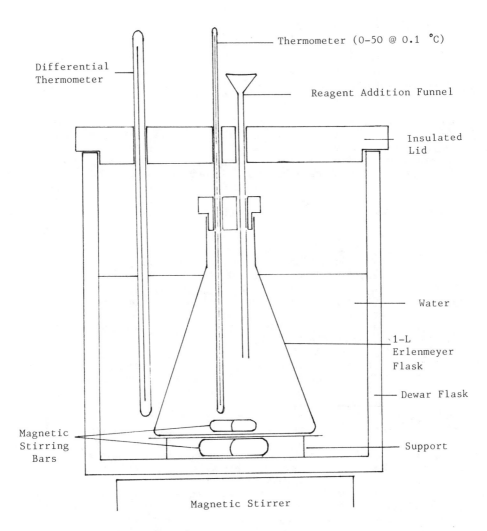

Figure 8. Simple calorimeter.

The Environmental Coordinator is responsible for our compliance
with laws regarding effluent and air emissions. He informs us of
any problems such as environmentally undesirable solvents or
reagents. He also determines the best way to dispose of any solid
or liquid waste streams resulting from a procedure. If required, he
arranges testing for a definitive answer to questions of appropriate
disposal.

The Safety Officer, after receiving the R & D process description, immediately consults the appropriate literature to determine
if any hazard exists relative to the toxicity or irritability of a
particular product or intermediate. Material Safety Data Sheets,
when available, are his prime source of information. In their
absence, he consults the supplier of either the material or process
for further information. If no information is available, he submits
samples for toxicity screening, if it appears warranted. He may
arrange testing for irritation, mutagenicity (Ames Test) or other
hazards. Once all chemical toxicity information is available, he is
responsible for judging its relative severity. He also recommends
suitable protective equipment to be used by manufacturing personnel
to avoid contact with a potentially hazardous substance.

Hazard Review Summary

After all individuals complete their respective examinations, a
meeting of the entire Hazard Review Team is held. One of the
chemical operators from the Pilot Plant also attends this meeting.
We have found that the presence of an operator provides personnel's
overall reaction to a process' ease of handling and general acceptance in the plant. Good public relations is also involved when it
becomes known how much effort is put into making a process run
safely.

Although no strict format has been developed for these meetings,
they usually follow the same general course. The meeting begins with
individual members of the Team summarizing their reports (safety,
toxicity, reactivity, etc.). This is followed by a very detailed
examination of the process itself where various technical points are
clarified (e.g., exactly how is a particular transfer going to be
carried out, what reactors will be used, what type of filtration,
etc.). During this discussion Team members interrupt with additional
handling and safety details. By the time the process has been
summarized, a few obvious items for further discussion have been
targeted. When these points have been resolved to everyone's
satisfaction, the Hazard Assessment Form is circulated for any
signatures not already present. The meeting is then adjourned.

From the notes of this meeting the Pilot Plant Director
prepares the Hazard Review Summary. This Summary specifies the
necessary engineering controls, work practices, administrative
procedures and personal protective equipment necessary to manufacture the product at a low level of risk. This Summary is
circulated to appropriate personnel.

As the last stage of the hazard review process, the R & D
Chemist prepares and arranges for the issue of a finalized manufacturing monograph for use during piloting. This Batch Record includes
all appropriate warnings and cautions and is written in a form that
is usable for regular factory production. A copy of this Batch

Record, as well as a completed Hazard Assessment Form, must be on file with the Pilot Plant Director before any pilot operations can begin.

If a product is to be scheduled for subsequent factory production, a meeting of the Hazard Review Team is held again. At this point the Team also includes the Area Supervisor and Foreman who are involved in the project. Any unusual, unexpected, or pertinent facts gleaned from the piloting experience are discussed and any necessary process modifications are clarified.

The effectiveness of any hazard avoidance program is, of course, directly dependent upon the training and safety consciousness of the operators implementing the process. We have addressed this question for years with an involved training program which covers in detail the entire spectrum of different procedures the operator will be expected to carry out.

Safety Analysis Worksheets were compiled to contain the safest work practices for a given situation. These Sheets, on file in the Pilot Plant, describe the potential hazards possible in a given situation (e.g., adding charcoal to a hot solution) and outline the exact precautions to be exercised in performing such operations.

Since Pilot Operators are expected to carry out a much wider variety of operations than would be an operator in the factory, the possibility exists that one may get rusty where a particular operation is concerned. To overcome this problem, we have developed the Piloting Techniques Worksheet. These Worksheets describe exact methods and techniques for performing operations in the Pilot Plant. They also are immediately accessible in the building.

In closing, I would like to emphasize that hazard avoidance at Rensselaer is considered a serious and ongoing effort. If at any time a question arises concerning the safety of a process during piloting, we reconvene the Hazard Review Team and further identify problems and solutions. In retrospect, we have found that most of our problems arise from trying to work too fast (e.g., without a Batch Record). The hazard review process has done much to make our operation a safer one.

Acknowledgment

The author thanks Dr. Colin F. Coates (Sterling Organics Ltd., U.K.) for his help in developing and implementing our hazard evaluation program.

Literature Cited

1. Figures 1 - 3: Coates, C. F. In "A System of Hazard Evaluation for a Medium-Sized Manufacturer of Bulk Organic Chemicals"; I. CHEM. E. SYMPOSIUM SERIES NO. 68, The Institution of Chemical Engineers: Rugby, England, 1981; p. 4/Y:1, 4/Y:13.

RECEIVED November 3, 1984

Thermochemical Hazard Evaluation

ROBERT C. DUVAL

Chemical Development Section, Sandoz Research Institute, Sandoz, Inc., East Hanover, NJ 07936

Thermochemical hazard evaluation should be an integral
part of any chemical process hazard review. This paper
discusses how a pharmaceutical chemical development
group performs thermochemical hazard evaluation through
a combination of literature searches and physical
testing. It will briefly discuss the physical testing
methods and some of the philosophy behind them.

The evaluation of potential thermochemical hazards should be an
integral part of any chemical process hazard review. An uncontrolled release of heat during a chemical process operation can
lead to problems that can vary from an inconvenient loss of a
product batch to a devastating explosion.

Thermochemical hazards are only one of many types of potential hazards in chemical processes, but they are of special concern because they are not easily identified and assessed. Even
after it is determined that a potential thermal hazard exists, it
must be decided whether that hazard can be avoided, controlled, or
accepted. Identification of thermochemical hazards becomes even
more of a problem when the exact composition of the material being
handled, such as a distillation residue, is unknown, or when
seemingly unimportant factors, such as catalytic amounts of
impurities in starting materials, play an important part in the
thermal stability of the process material.

The best way to evaluate thermochemical hazards will vary
from one laboratory or plant situation to another depending on
such factors as the stage of process development, the size of
scale-up necessary, the equipment available, the time available
for hazards review, and the amount of risk acceptable.

Our Chemical Development group is involved in process
development and production of pharmaceutical compounds. The bulk
of the work involved is in the early stages of process development
with the purpose of quickly supplying our chemistry, pharmacology,
toxicology, pharmaceutical development, and clinical research

0097-6156/85/0274-0057$06.00/0
© 1985 American Chemical Society

groups with material for their studies. These material demands can vary from 0.3 to 15.0 kilograms. Their production involves the use of 5 to 22 liter flasks in labs for small requirements or 50 to 1000 liter reactors in development plants for larger requirements. Many of the processes will be scaled-up only one or two times because of the high dropout rate of pharmaceutical research compounds.

Our thermochemical hazard evaluation process begins as soon as a project or procedure is received by Chemical Development. We are trying to improve safety not only for the plant personnel but also for the process research chemists who will be developing the process in the lab.

Literature Search

Our evaluation utilizes a combination of literature searches and physical testing. The literature search is a part of a comprehensive literature search for all types of biological and chemical hazards. It is performed by the chemist in charge of the project before any other work is started, and it is updated to include any changes as the process is developed. The literature sources used in these searches that are of interest for the identification of thermochemical hazards are listed in Table I. The Sandoz Ltd., Chemical Development Safety Data File is a computer data file that contains safety information from proprietary and published sources.

Table I.   Literature Sources

---

1. National Fire Protection Association, Fire Protection Guide on Hazardous Materials 7th Edition, Boston, MA, 1981.
2. Sax, N. Irving, Dangerous Properties of Industrial Materials 5th Edition, New York, Van Nostrand Reinhold Company, 1979.
3. Bretherick, L., Handbook of Reactive Chemical Hazards 2nd Edition, Cleveland, Ohio, CRC Press, Inc., 1979.
4. Chemical Abstracts Computer Search.
5. Sandoz, Ltd., Chemical Development Safety Data File Computer Search.

Thermal Stability Testing

Our physical testing program is concerned with two main areas, thermal stability and reaction calorimetry. The thermal stability testing is broken down into two phases, initial screen and follow-up tests. The initial screen is intended to quickly identify any thermally unstable materials in a process. The follow-up tests examine in more detail any significant instability detected in the initial screen.

The types of samples that we test for thermal stability are starting materials, isolated intermediates, evaporation residues, distillation residues, products, and evaporated mother liquors. The evaporated residues refer both to complete and partial reaction concentrations. We also test reaction mixtures and

non-isolated intermediates depending on the process or the test
results of other related samples.

Because of our early involvement in process development, we
have the limitations of short time, limited sample size and a
large number of samples. These limitations cause us to use tests
that tend to be more qualitative than quantitative.

Initial Screen. Our initial screen is summarized in Table II. It
consists of two test methods, Differential Scanning Calorimetry
(DSC) and Differential Thermal Analysis (DTA). In both of these
methods the sample is exposed to heat, and thermal changes in the
sample are recorded. Heat flow into and out of the sample is
recorded in DSC, and the temperature difference between the sample
and a reference is recorded in DTA.

Table II. Initial Screen

Test	Sample Size	Safety Margin
Differential Scanning Calorimetry		
Dynamic (10°C/min., 0 & 500 psig)	1-3 mg	50°C
Differential Thermal Analysis		
Dynamic (2.5°C/min.)	2-5 g	100°C
Isothermal (8-20 hr.)	2-5 g	50°C

For the DSC tests we use a Dupont 1090 with the pressure DSC
measuring cell. In our DSC tests we use a dynamic heating method
with a heating rate of 10°C/min. We run two tests, one at
atmospheric pressure and one at 500 psi of applied pressure.

For our DTA tests we use equipment from Adolf Kuhner AG (1).
We run two types of DTA tests. One type is a dynamic heating
method with a 2.5°C/min. heating rate, and the other is an
isothermal heating method. In the isothermal test we preheat the
heating block to a specific temperature, insert the sample, and
keep the heating block at the specific temperature for at least
eight hours.

In all of the initial tests we are interested in finding out
if the sample shows any thermal instability, and if so, at what
temperature it is first detectable. We are most concerned with
thermal instability in which heat is released (an exotherm).
Thermal instability in which heat is absorbed (an endotherm)
should not be ignored, however, because it may represent significant gas evolution.

Dynamic heating methods are used in the initial screen for
two reasons. First, they allow a quick test over a large temperature range. Second, they are sensitive in detecting thermal
transitions, as illustated in Figure 1. This is a comparison of
DSC curves of $\underline{N},\underline{N}$-Diphenylhydrazine hydrochloride in which the
sample has been heated at different rates. The curves show an
exotherm that becomes sharper and apparently larger with increasing heating rates. The peak intergration values are included to
show that more heat is not really being evolved with increased

heating rates, but that the same amount of heat is being evolved
in a shorter period of time.   This peak sharpening that occurs
with dynamic heating causes exotherms to be more readily observed,
and therefore, the method is more sensitive in detecting smaller
exotherms.

Figure 1 also illustrates why a dynamic heating method is not
suitable for obtaining information on the lowest temperature at
which thermal instability can be detected.   The curves show
initial baseline deflections that shift to higher temperatures
with increased heating rates.   With the constantly changing
applied heat of a dynamic test, there will be varying delays in
initial transition detection because of sample and instrument
response lags.   Isothermal heating methods are necessary to provide more significant initial detection temperatures.

Another reason that isothermal heating methods are used in
the initial screen is to identify materials that have time dependent thermal stability.   These materials have a thermal decomposition that does not follow a simple Arrhenius relationship in which
the reaction rate increases exponentially with an increase in
temperature.   Instead an extended induction period is required
before the decomposition becomes detectable.   An example of this
behavior is shown in Figure 2.   The DTA isothermal test recorder
traces of methane sulfonic acid, 3,7-dimethyloctyl ester at
different test temperatures are shown.   The induction time varies
from less than 1 hr. at 180°C to 46 hr. at 130°C.   As with this
compound, it is not unusual that once decomposition is detected it
proceeds very rapidly, releasing all of the heat in a short period
of time.   Dynamic heating methods do not indicate if this type of
thermal instability is present; if it is, the initial detection
temperature from dynamic tests will be grossly misleading as to
the thermal stability of the material.

The initial screen uses both DSC and DTA dynamic heating
method tests to compensate for some of the problems inherent in
each test.   The DSC test is fast, simple, sensitive, and quantitative.   It requires only a small amount of sample.   The small
sample size, however, can be a problem with some samples, such as
distillation residues, because of a lack of sample homogeneity.
Also, the only inexpensive sample containers for DSC are aluminum
or stainless steel.   The containers can sometimes cause problems
because of chemical reactions between the sample and pan.   The DTA
test addresses both of these problems.   It uses a 2-5 g sample and
the containers are glass.   Its disadvantages are that it lacks the
sensitivity of the DSC and that it is not quantitative.

A pressure DSC test (semi-closed sample pan, encapsulated in
air, under 500 psig of N_2) and a non-pressure DSC test (semi-
closed sample pan, encapsulated in air, at ambient pressure) are
used in our initial screen for several reasons.   The pressure DSC
allows the thermal stability of liquids to be examined near, at,
or above their boiling points.   It also suppresses the evaporation
of volatile materials from the sample, which can hide an exotherm.   This can be especially important when testing evaporation
residues.   A comparison of the results of both tests gives an
indication of the effect of pressure on the decomposition of the
material.   It will also give an indication if oxidation is an

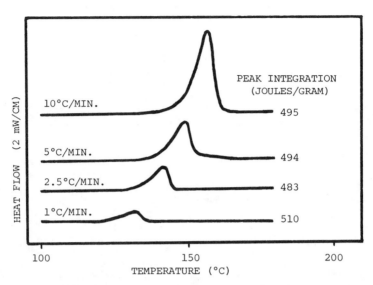

Figure 1.   DSC curves of N̲,N̲-Diphenylhydrazine hydrochloride at different heating rates.

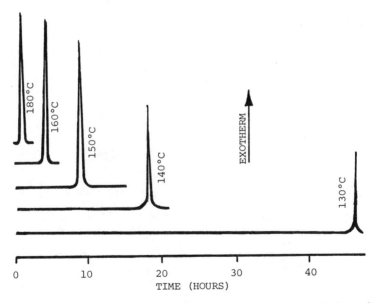

Figure 2.   DTA recorder traces of a compound which exhibits time dependent thermal stability.

important factor. Finally, the comparison helps us decide if semi-closed or closed containers would be more appropriate for our DTA tests than the usual open container. A more detailed discussion of the use of pressure DSC in thermal hazard evaluation can be found in an article by R. J. Seyler (2).

During our initial screen, we also check the thermal stability of the sample in the presence of stainless steel. Accidents have occurred when processes that were no problem in glass equipment were either scaled-up in, or switched to stainless steel equipment. We perform this check during our isothermal DTA test by running duplicate samples and adding stainless steel powder to one of them. We use the isothermal test because it allows longer contact time between the sample and the stainless steel.

If our initial screen detects some thermal instability, we need some way to decide if the instability represents a possible hazard in the process. We do this by comparing the lowest temperature at which we detect a sample's instability in each test with the sample's highest process exposure temperature. If these temperature differences fall within predetermined safety margins for any one or more of the tests, we will examine the thermal stability of the material further. These margins are based on our experience as to how much the detection temperature of the instability can be lowered in our follow-up tests. For example, if we were to detect an exotherm in the DSC test higher than 50°C above the process temperature, and this exotherm was also detected outside the other test safety margins, further detailed testing would usually not lower this detection temperature to a point where we would consider the exotherm to be an unacceptable hazard. These temperature margins are only guidelines and can vary according to the process under review or the test results themselves. Two examples of when we might expand the temperature safety margins are when the initial tests detect large exotherms near the margins or when the sample is known to be a serious potential hazard, such as an aromatic nitro compound.

Follow-up Tests. In our follow-up tests we want to better define at what temperature we can detect the thermal instability, and to gain some knowledge about how much of a hazard the instability represents. The majority of this work is done with an instrument called a Sikarex Safety Calorimeter (3,4). It consists of a sample oven, a control and measurement module, and a recorder. In the sample oven is placed 10-30 g of sample in either an open glass tube, a closed glass tube with a capillary bleed, or a stainless steel autoclave. The control and measurement module controls the oven temperature and measures the sample and oven jacket temperatures.

We run two types of tests on the Sikarex. The first involves step-heating the sample through a temperature range which is determined from the results of the initial screen. We usually elevate the jacket (oven) temperature in 10° or 20°C increments, and hold the jacket temperature constant for 1-2 hr. after the sample and jacket temperature have equilibrated. We analyze the data by plotting the jacket temperature versus the temperature difference between the sample and jacket (Figure 3). The upward deflection in the slope of the graph indicates an exotherm. The last point (A) on the baseline and the first point within the deflection (B) define the temperature range in which the exotherm is first detected. This temperature range is what we consider to be the lowest significant exotherm detection temperature. It is usually 10-50°C lower than the detection temperatures we obtain in the initial screen.

The second type of test we run on the Sikarex is an adiabatic test. In this test the jacket temperature is controlled by the sample temperature. When the sample thermometer detects an increase in the sample's temperature, the jacket temperature is increased an equal amount. In other words, the sample is being held under adiabatic conditions. The test is run by step-heating the sample into the exotherm detection range found in the previous Sikarex test by means of a heating coil attached to the sample tube. External heating is then stopped, and the sample is allowed to self-heat. The adiabatic temperature rise of both the jacket and sample are recorded.

Figure 4 illustrates two possible types of results from the adiabatic test. The circles show an exotherm with a large adiabatic temperature rise and a rapid self-heating rate. This test result would indicate a high hazard potential associated with the exotherm and a hazard we would want to avoid. The triangles show an exotherm with a small adiabatic temperature rise and slow self-heating rate. This test result would indicate a low hazard potential associated with the exotherm and a hazard that would be of less concern as far as its ability to cause a serious accident.

The adiabatic Sikarex test results also give us an idea of how long the decomposition reaction takes to reach its maximum reaction rate. We do not attach significance to the exact length of time, but we use it as an indication of whether the time to maximum rate is short (minutes) or long (several hours or days).

The adiabatic tests give us some idea of the hazard potential associated with an exotherm which helps us to decide the extent of avoidance or precautions that are necessary in the procedure. We generally like to have at least a 20°C temperature margin between the Sikarex isothermal exotherm detection temperature range and the highest process exposure temperature. This margin will increase in cases where the adiabatic test shows a high hazard potential, and it will possibly shrink if the adiabatic test shows a very low hazard potential. Again, these margins are generalizations, and they will vary depending on the process and on other test results.

In our follow-up testing we also run other special types of tests to further examine possible hazards identified in our initial screen or to clarify the significance of Sikarex test results for our process. Some of these special tests are listed in Table III.

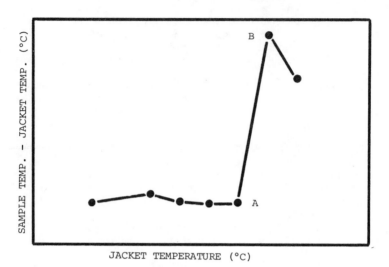

Figure 3.   Example of Sikarex step isothermal data.

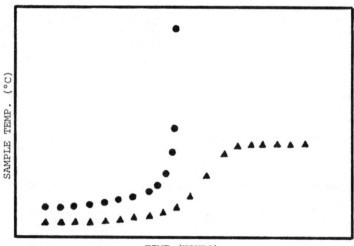

Figure 4.   Example of two possible types of Sikarex adiabatic
test results.

Table III: Special Tests

Type of Study	Instrument
Effect of pressure	DSC
Effect of different atmospheres	DSC
Heat of decomposition	DSC
Kinetic approximations	DSC
Pressure generation	DTA, Sikarex (autoclaves)
Evolved gas - volume/rate	DTA, Sikarex
Effect of solvent or other materials	DSC, DTA, Sikarex
Extended isothermal testing	DSC, DTA, Sikarex
Reaction calorimetry	Reaction Calorimeter

Reaction Calorimetry

The second main area of our physical testing program is reaction calorimetry. It is an extremely useful test method for identifying and assessing the hazards associated with running exothermic reactions (4,5). The control of an exothermic reaction can be a serious problem if you have not properly designed your procedure or chosen inappropriate process equipment. For example, insufficient mixing speed or incorrect reaction temperatures can lead to an accumulation of reagents which can then react uncontrollably. Adding reagents too quickly or heating too rapidly can also lead to uncontrollable reaction rates. Reaction calorimetry involves running the reaction according to the process procedure and measuring the heat changes of the reaction mixture as the reaction proceeds.

In order to carry out these measurements, we use a reaction calorimeter that was designed by Dr. L. Hub and his group in our Chemical Development Safety Lab at Sandoz Ltd., in Switzerland. It consists of a one liter reaction vessel along with the necessary equipment for temperature control and quantitative measurement of heat flow into and out of the reaction vessel.

In this test we are usually most interested in obtaining the rate and amount of heat released. We usually analyze the results by using graphs. One example is shown in Figure 5. This data was obtained from an exothermic oxidation reaction in which hydrogen peroxide was added to the reaction mixture. There was concern about possible reagent accumulation due to improper addition rates. The measured heat evolution rate and the hydrogen peroxide addition rate have been plotted together versus time. A profile of the heat released in relation to the amount of reagent added is obtained. Integration of the heat evolution curve gives the total heat of reaction.

We would like to have reaction calorimetry data for every reaction we scale-up. However, due to time and capacity constraints, we only run calorimetric measurements on selected reactions that we feel show the greatest potential to cause a problem. Some examples are: reactions that have been known to cause problems in the past (our own experience or literature), reactions that show potential problems during lab scale development work

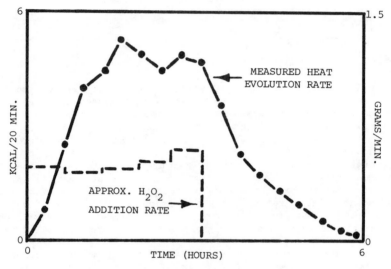

Figure 5. Example of results from a reaction calorimetry measurement.

(delayed exotherms, control problems), reactions that have all the reagents mixed together before heat is applied, and reactions that contain materials our thermal stability tests show to be potential hazards.

Closing Comments

There are several points to be kept in mind when using physical testing as part of process hazard evaluation. First, the limitations of the test method should always be kept in mind. For example, it has been pointed out that different thermal stability tests give different exotherm detection temperatures. In most cases it is not possible to define an exact exotherm onset because the decomposition reaction's rate does not go to zero as the temperature is lowered. Overconfidence in test results can be just as much of a hazard as no knowledge at all if the limitations of the tests are forgotten.

The second point is to always run tests on representative samples. Table IV illustrates this point. Original thermal stability tests were run on an alpha-oximino ester intermediate product that had been isolated by adding water to the reaction mixture, extracting the oil layer that forms with methylene chloride, and removing the methylene chloride by vacuum distillation (labelled pure oil). Later in process development it was decided to eliminate the methylene chloride extraction and separate the oil from the water layer (labelled crude oil). If repeat safety tests had not been run, the thermal stability hazard of this compound might never have been realized, and the compound might have been improperly stored or handled at too high a temperature.

Table IV: Test Results of an Alpha-Oximino Ester Intermediate

Test - Type of Results	Results	
	Pure Oil	Crude Oil
DTA Dynamic - exotherm detection temp.	140°C	65°C
Sikarex Step Isothermal - exotherm detection range	100-120°C	20-30°C
Sikarex Adiabatic - adiabatic temp. increase / time interval	120 to 270°C* in 2.5 hr.	40 to 200°C* in 5.5 hr.

* Sudden gas evolution and expulsion of material from the test tube.

The third point is to consider and review all test results in the context of the procedure under evaluation. Safety test results are not like NMR or melting point data. In order to be useful they must be evaluated with the process and equipment in mind.

The final point is to transfer the test results to the people who can use them. Not only are the safety test results important during process planning, they are also important during the actual process scale-ups and production. The people who are running the process need to be aware of the test results and have them available so that they can make educated decisions if unexpected process changes are necessary or if emergencies arise.

If the results of the literature searches and physical tests detect a hazard that we feel is unacceptable, we will usually try to avoid the hazard by changing process conditions and including written warnings in the procedure. If this is not possible, we will try to control the hazard by adding safety precautions to the procedure and changing the equipment to run the process. If avoidance or control still does not lower the risk to a level we feel is acceptable, we will reject the procedure and look for a different way to produce the product.

The most important consideration for avoiding hazards is to recognize them. By identifying a potentially dangerous situation and analyzing that situation in a manner appropriate to the operation, the chances of having an accident are significantly reduced.

Literature Cited

1. Adolf Kuhner AG, Apparatebau, CH-4052 Basel, Switzerland. Equipment for Safety Test According to Ciba-Geigy-Kuhner.
2. Seyler, R. J., "Application of Pressure DTA(DSC) to Thermal Hazard Evaluation", Thermochimia Acta, Vol. 39, 1980, p. 171-180.
3. System-Technik AG, CH-8803 Ruschlikon, Switzerland.
4. Hub, L., "Two Calorimetric Methods for Investigating Dangerous Reactions", Chem. E. Symposium Series No. 49, 1977.
5. Giger, G. and Regenass, W., "Assessment of Reaction Hazards by Means of a Bench Scale Heat Flow Calorimeter", Proc. Eleventh North American Thermal Analysis Society Conf., Vol. 2, pp. 579-586, 1981.

RECEIVED November 3, 1984

Thermal Runaway Reactions: Hazard Evaluation

LINDA VAN ROEKEL

Columbia Scientific Industries, Austin, TX 78759

An investigation of potential thermal runaway
reactions is a significant part of a thorough hazard
evaluation. Important parameters of the exothermic
reaction as well as of the large-scale system are
discussed. Their relationship is explained through
the Semenov Theory.

In reviewing the hazard associated with a chemical process, one of
the hazards which should be considered is that of a potential runaway reaction. If either the desired chemical reaction or an undesired reaction (e.g., a side reaction or the unintended decomposition of a product) produces more heat than can be dissipated, the
heat will accumulate in the system. This can lead to the thermal
runaway. If the exothermic reaction(s) is accompanied by significant
pressure generation, the runaway reaction can lead to rupture of the
reaction vessel.

In order to study the potential for a runaway reaction, the
investigator must be aware of the characteristics of the chemical re-
action(s) as well as the characteristics of the actual large-scale
system. In other words, a review of the hazards of an exothermic
reaction requires a knowledge of both the "chemistry" of the reaction
and the "engineering" of the large-scale system.

The "Chemistry" of the Exothermic Reaction

For the thermal runaway hazard evaluation, the "chemistry" of the
exothermic reaction can be defined in terms of three sets of parameters: the thermodynamic, kinetic, and physical parameters. (1) A
list of some of the parameters of interest is given in Table I.

0097-6156/85/0274-0069$06.00/0
© 1985 American Chemical Society

Table I. Parameters to Define the Exothermic Reaction

---

1. Thermodynamic Parameters
   * Adiabatic Temperature Rise
   * Reaction Energy
   * Moles of Gas Generated
   * Maximum Pressure in a Closed Vessel
2. Kinetic Parameters
   * Reaction Rate
   * Rate of Heat Production
   * Rate of Pressure Generation
   * Adiabatic Time to Maximum Rate
   * Apparent Activation Energy
   * Detectable Onset Temperature of Exotherm
3. Physical Parameters
   * Heat Capacity
   * Thermal Conductivity

Source: Reproduced with permission from Ref. 1. Copyright 1982
Chem. Eng.

Thermodynamic Parameters. The adiabatic temperature rise, ΔT_{AB} is
the temperature rise associated with a given reaction if that reaction is run under conditions of no heat transfer. This temperature
rise is directly proportional to the heat of reaction through the
relationship

$$\Delta H = C_p \times \Delta T_{AB} \tag{1}$$

The change in enthalpy or the heat of reaction is the amount of heat
released during the exothermic reaction, but one should also be aware
of the reaction energy, ΔE, which is $\Delta H - \Delta(PV)$. Most industrial
processes are constant volume processes so the reaction energy takes
into account the change in pressure for such processes.
    The last two thermodynamic parameters listed also deal with the
pressure generated. The moles of gas generated per unit of reaction
mass, along with the void space of the container, will be used in determining the maximum pressure which will be reached in the closed
vessel. The pressure measurements are significant since the pressure
and the integrity of the container will determine the potential for
rupture of the container.

Kinetic Parameters. Not only does the investigator need to know how
much heat and how much pressure are generated but also how fast they
are being generated. The reaction rate is the conventional means of
expressing this. For an nth order reaction involving a single reactant, the rate of reaction is usually given as the rate of disap-
pearance of the reactant or

$$\frac{-dC}{dt} = kC^n \tag{2}$$

For some simple reactions, the rate constant, k, can be expressed by the classical Arrhenius equation:

$$k = A \exp (-E_a/RT) \tag{3}$$

where A is the pre-exponential factor, E_a is the activation energy, and R is the universal gas constant.

For a hazard review, we are specifically interested in the rate of heat production and the rate of pressure generation. The rate of heat production (temperature rate or self-heating rate) and the rate of pressure generation depend upon the temperature and the degree of conversion. In some instances the self-heating rate may also depend upon the thermal history of the material. This is true, for example, with autocatalytic reactions.

The adiabatic time to maximum rate, TMR, gives a measure of the time required to reach, from a given temperature, the maximum self-heating rate for a system under conditions of no heat transfer. A plot of TMR vs. temperature is shown in Figure 1 for the decomposition of di-tert-butyl peroxide. The time to maximum rate is best measured directly rather than calculated because of the very large errors associated with the exponential term involved in the calculations. (2) TMR can be measured directly using an adiabatic calorimeter such as the Accelerating Rate Calorimeter.

For simple, single reactions, it is often possible to determine the Arrhenius activation energy. For complex systems, sophisticated modeling techniques may give an apparent activation energy.

The final kinetic parameter listed in Table I is the detectable onset temperature of the exotherm. The adjective "detectable" is extremely important. The measured onset temperature of an exotherm is instrument dependent. For a non-autocatalytic, uninhibited decomposition for example, the adiabatic course of the reaction can be represented by plotting the logarithm of the self-heating rate vs. 1/T. A typical plot is shown in Figure 2. If the measuring technique detects an exotherm at a rate of 1°/minute, the onset temperature would be measured here as about 140°. The Accelerating Rate Calorimeter detects an exotherm at 0.02°/minute (3) and would detect this reaction at 100°. An instrument or technique which is even more sensitive than the ARC would find an even lower detectable onset temperature. Under truly adiabatic conditions (no heat loss), any heat generation will lead to a rise in temperature which will then lead to higher self-heating rates and so on. The time required for the reaction to generate a "significant" amount of heat or pressure (from a given starting temperature) is a measure of the safety of the system.

Physical Parameters. Both the heat capacity and the thermal conductivity play a role in the calculations which need to be made. Materials with poor thermal conductivity, e.g. solids, are difficult to evaluate in terms of thermal hazards, because the poor thermal conductivity can contribute to the development of "hot spots."

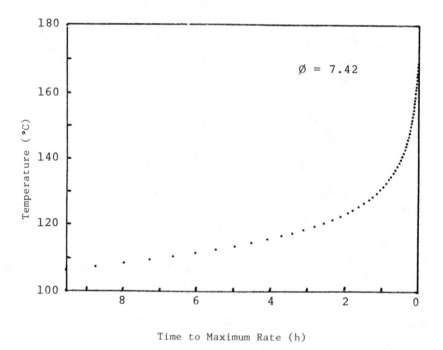

Figure 1.   Temperature vs. Time to Maximum Decomposition Rate
(TMR) for Di-tert-butyl Peroxide.

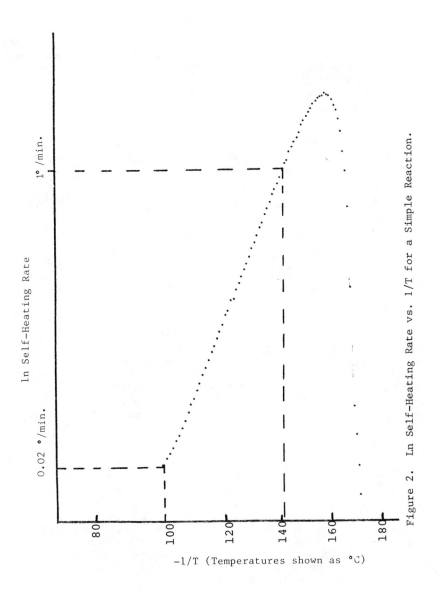

Figure 2.   Ln Self-Heating Rate vs. 1/T for a Simple Reaction.

The Engineering of the Large-Scale System

Some of the characteristics of the large-scale system which are of
interest in a thermal runaway hazard evaluation are listed in
Table II. First of all, the amount of material which will be handled must be known. The hazards involved are obviously greater when
one works with large quantities of material. The heat generated on
a laboratory scale is so much less because of the small quantities
of material being handled. Often that heat is easily dissipated because of the low sample volume to surface area ratio. Secondly, the
heat transfer characteristics of the system must be known. Is
special cooling available? What is the surface area through which
heat can be dissipated?
    Only batch processes will be considered here. Continuous
processes have the added safety advantage of continually removing
products (and heat) from the system.

Table II.   Parameters to Define the Large-Scale System

1.   Amount of Material
2.   Heat Transfer Characteristics
3.   Batch vs. Continuous

Semenov Theory

For systems with a uniform temperature throughout the material, the
"chemistry" and the "engineering" can be related through use of the
Semenov Theory. (4) The rate of heat production was mentioned
earlier as a kinetic parameter of interest and the heat transfer
characteristics (in this case, rate of heat removal) as a large-
scale system parameter. If the self-heating rate (rate of heat production) is determined as a function of temperature under adiabatic
conditions and if there is a knowledge of the rate of heat removal
as a function of temperature, information about safe operating limits
for that particular system can be deduced.
    In Figure 3, the curved line represents the heat generation
rate (self-heating rate) as a function of temperature under adiabatic
conditions. That is, under adiabatic conditions or no heat transfer,
heat will be generated (in this hypothetical reaction) according to
the function shown. In the actual chemical process, however, some
heat will be removed. The straight line represents the rate of heat
removal as a function of temperature. The slope of this heat removal line is U x S where U is the heat transfer coefficient and S
is the surface area through which heat can be dissipated. The
intercept of the line with the x-axis is the temperature of the
coolant, T_0.
    For the system as represented in Figure 3, there are two points
of intersection or two points at which the system is in equilibrium.
At these points, the rate of heat generation is exactly counter-
balanced by the rate of heat removal. Point A is a steady state

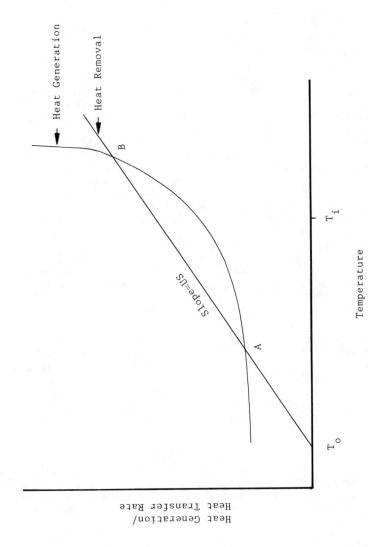

Figure 3.   Semenov Plot for a System Which Exhibits a Steady
State Condition (at A).

condition. Suppose that an upset condition causes an increase in temperature to T_1. When normal conditions are restored at T_1 the rate of heat removal will be greater than the rate of heat generation. The reaction mass will slowly return to the conditions described at Point A. Point B at the higher temperature is actually a meta-stable state since a slight perturbation of the system from these conditions can result in a rate of heat generation greater than the rate of heat removal and a runaway situation.

Figure 4 illustrates an unsteady state condition. At all temperatures, the rate of heat generation is greater than the rate of heat removal. There is no equilibrium state. This situation will result in a runaway.

Figure 5 shows the third possibility for the relative positions of the heat generation and heat removal lines. This represents the critical state. That is, there is only one point of intersection between the two curves, only one point at which the heat removal rate is exactly equal to the heat generation rate. This is a point of equilibrium, but if the rate of heat generation should increase (through an impurity which acts as a catalyst, for example), or the rate of heat removal should decrease (e.g., through scale build-up or an increase in the temperature of the coolant), a runaway situation will occur. This critical point is often referred to as the Temperature of No Return, T_{NR}. (5) Note that this is the temperature of the reaction mass and not the temperature of the coolant. T_0' is the temperature of the coolant under these conditions with the difference between the temperatures of the reaction mass and the coolant being ΔT_{CR}.

It has been shown by Townsend and Tou (5) that from T_{NR} the time to maximum rate, $(\theta_{MR})_{T_{NR}}$ can be calculated from the equation:

$$(\theta_{MR})_{T_{NR}} = M \times C_p / U \times S \qquad (4)$$

It should be pointed out that the Semenov Theory was developed for gases, is generally applied to non-viscous liquids, but does not hold for solids. Solids will not show a uniform temperature distribution because of their poor thermal conductivity. For solids, a more complex model must be used, such as the Frank-Kamenetskii Theory. (6) Discussions of this theory and others can be found in the literature. (7)

Critical Parameters

T_{NR} is a critical temperature in the sense that it is the highest allowable temperature for a material under given conditions of heat generation and heat transfer. Likewise, one can calculate a critical radius, r_{CR}, and a critical volume, V_{CR}. The critical radius and critical volume are the largest values of the respective parameter for which the heat generated can still be safely dissipated. Table III (1) illustrates the major changes in r_{CR} and V_{CR} for a (relatively) small change in the temperature. Note, for example, that the critical radius decreases from more than 8 meters to about 28 cm when the temperature of the material is increased from 100° to 120°. Correspondingly, the time to maximum rate, TMR,

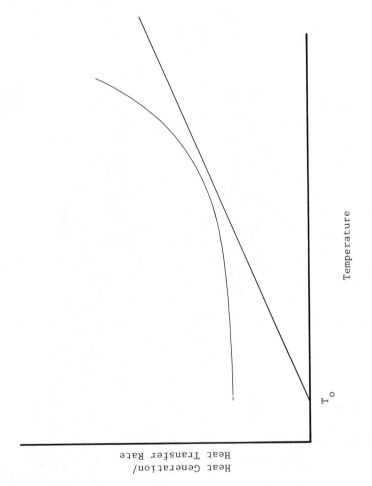

Figure 4.   Semenov Plot for a System in an Unsteady State.

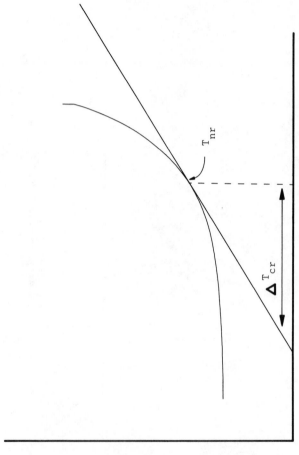

Figure 5. Semenov Plot for a System with only one Point of Equilibrium, the Critical State.

from 100° to 120° decreases from about 1 1/2 days to approximately
3 hrs.

Table III. Adiabatic Time to Max Rate and Critical Volumes and Radii

Temperature, $^\circ$C	TMR	r_{CR}	V_{CR}
75	61.3 days	1.26×10^4 cm	2.38×10^7 m^3
100	38.5 hrs.	830 cm	225 m^3
125	3.25 hrs.	27.9 cm	170 cm^3

For di-tert-butyl peroxide in a cylinder with h = 2r and

$$A = 10^{15} \text{ sec}^{-1}$$
$$\Delta T_{AB} = 500 \ ^\circ K$$
$$E_a = 156.9 \text{ kJ/Mole}$$
$$\rho = 0.9 \text{ g/cm}$$
$$C_p = 2.1 \text{ J/g } ^\circ K$$
$$U = 1.5 \times 10^{-3} \text{ J/cm}^2 \ ^\circ K \text{ sec}$$

Source: Reproduced with permission from Ref. 1. Copyright 1982
Chem. Eng.

Cautions

In conducting an investigation of thermal hazards, particularly of
the thermal runaway, certain cautions must be mentioned. As discussed previously, the Semenov Theory holds for many liquids, but
solids must be treated quite differently. Because the Semenov Theory
is easier to apply than the theories available for solids, it is
often tempting to apply the Semenov Theory to solids as well as
liquids. Major errors can arise!
     Also, beware of autocatalytic reactions. The examples given in
this paper are for non-autocatalytic, uninhibited systems. Autocatalytic or inhibited materials will exhibit different thermal
behavior depending on their thermal history. Just being able to recognize these materials is an important part of a hazard evaluation.
If such materials are thermally aged, they will show a lower detectable onset temperature than fresh, untreated material.
     The equations presented in this paper have assumed Arrhenius
kinetics. Many chemical reactions do not proceed according to
Arrhenius kinetics.
     Finally, determination of the potential for a thermal runaway
is only one part of a thorough hazard evaluation. Flammability,
potential for dust explosions and shock sensitivity are only a few
of the other hazards which may lurk in every chemical process.

Literature Cited

1. Smith, D.W. Chemical Engineering December 13, 1982, 79-84.
2. Wilberforce. J.K. J. Thermal Analysis 1982, 25, 593-6.
3. Smith, D.W.; Taylor, M.C.; Young, R.; Stephens, T. American Laboratory June, 1980.
4. Semenov, N.N. "Some Problems of Chemical Kinetics and Reactivity"; translated by J.E.S. Bradley, Pergamon Press: Elmsford, N.H., 1959; Vol 2.
5. Townsend, D.I.; Tou, J.C. Thermochimica Acta 1980, 37, 1-30.
6. Frank-Kamenetskii, D.A. "Diffusion and Heat Transfer in Chemical Kinetics"; 2nd ed., translated by J.P. Appleton; Plenum Press: New York, 1969.
7. Davis, E.J.; Hampson, B.H.; Yates, B. In "Runaway Reactions, Unstable Products and Combustible Powders"; INST. CHEM. ENG. SYMPOSIUM SERIES No. 68, Institution of Chemical Engineers: Rugby, England, 1981, pp. 1/D:1-19.

RECEIVED November 3, 1984

The Thermochemical and Hazard Data of Chemicals
Estimation Using the ASTM CHETAH Program

CAROLE A. DAVIES[1], IRVING M. KIPNIS[2], MALCOLM W. CHASE[1], and
DALE N. TREWEEK[3]

[1] The Dow Chemical Company, Midland, MI 48640
[2] FMC Corporation, Baltimore, MD 21233
[3] 2626 Chartwell Road, Columbus, OH 43220

CHETAH, the ASTM Chemical Thermodynamic and Energy
Release Evaluation Program is a useful tool for the
preliminary screening of the reactivity hazard of
organic chemicals. This evaluation is based on
pattern recognition techniques with experimental
hazard data and estimated thermochemical data. Gas
phase thermochemical data for organic materials are
estimated with a second order group contribution
method. The use and capabilities of the CHETAH
program are reviewed and specific examples of its
application to Process Research and Development are
discussed.

CHETAH, which stands for Chemical Thermodynamic and Energy Release
Evaluation, is a computer program which can estimate thermodynamic
data and the potential reactivity hazard of chemical compounds (1).
CHETAH performs gas phase calculations over the temperature range of
25-1200°C. It contains a large data base of thermodynamic data,
which is used to calculate the energy release potential of a
compound or mixture. It is very useful for screening new products
because the only thing you need to know is the chemical structure;
therefore, you can evaluate the hazard before you attempt to make a
compound in the lab. The program calculates the maximum heat that
will be released if the compound decomposes. It does not estimate
the rate of the heat release. CHETAH is not intended to replace
physical testing, but only to be used as a screening tool to help
set priorities for physical testing as testing capacity is often
limited.

History

First, a brief history of the program and the people that developed
it is in order. American Society for Testing and Materials (ASTM)

0097-6156/85/0274-0081$06.00/0

Committee E-27, Hazard Potential of Chemicals, was formed in 1967, in response to the need for some agency to develop standardization techniques for the evaluation of the potential of chemicals to cause fires and explosions. Several subcommittees were formed and the one on Condensed Phase Reactions was given the assignment of devising a computational method for screening chemicals for their ability to cause an explosion. The task group that wrote the program, now known as CHETAH, was composed of Bill Seaton, Eli Freedman, and Dale Treweek. The original program was released in 1974. This task group has now formed a new subcommittee to deal specifically with estimation methods and is working on a new enhanced version of the program. The active members now include, in addition to the original authors, Ted Selover, Bob Alberty, Jim Fulton, Mal Chase, and Carol Davies.

Group Additivity Methods

CHETAH uses Benson's second-order group additivity method (2) to estimate the thermodynamic data. Most molecular properties of larger molecules can be considered, roughly, as being made up of additive contributions from the atoms or bonds in the molecule. The physical basis for this assumption is that the forces between the atoms in a molecule are very short range. Therefore, individual atoms seem to contribute nearly constant amounts to such molecular properties as entropy, heat capacity, and heat of formation. The simplest additivity method is a "zero-order". This assumes that the molecular property is the sum of all atomic contributions. A "first-order" additivity method is usually presented as a bond contribution method. It considers the structure of a molecule to some extent. The influence of the adjacent atoms on each other and on the property is taken into account. The "second-order" method developed by Sidney Benson uses contributions of groups, where a group is defined as a polyvalent central atom together with its attached ligands. Table I shows the results of the three methods for three compounds with the formula C_3H_8O. When we introduce structural features into a molecule which bring more distant parts of the molecule closer together, we can expect departures from the additivity laws. CHETAH contains ring corrections and next nearest neighbor corrections to compensate for this. As the order increases, the accuracy of the estimate of the thermal function increases. Also the magnitude of the necessary data increases by about a factor of ten for each order. By the time one has reached a second-order correlation for thermodynamic data, the estimates are almost as good as the data on which they are based, so there is little reason to go to any higher order.

TABLE I. Comparison of Additivity Methods for C_3H_8O Compounds

Molecule	ΔHf°_{298} (kcal/mol)	Estimation Method	Calculated ΔHf°_{298}	Error
$CH_3CH_2CH_2OH$	-61.55	0	-61.45	.1
		1	-60.35	1.2
		2	-60.13	.4
$CH_3-O-CH_2CH_3$	-51.73	0	-61.45	-9.7
		1	-51.91	-.2
		2	-51.58	.2
$\begin{array}{c} CH_3 \\ {\Large >} CH-OH \\ CH_3 \end{array}$	-65.15	0	-61.45	3.7
		1	-60.35	4.8
		2	-65.50	.4

An example of building the compound methallyl chloride with second-order groups follows.

$$CH_2{=}\overset{\overset{\displaystyle CH_3}{|}}{C}{-}CH_2Cl$$

Group	No. of Times Used
$C-(C_d)(H)_3$	1
$C_d-(H)_2$	1
$C_d-(C)_2$	1
$C-(C_d)(H)_2(Cl)$	1

The required groups are: a carbon attached to three hydrogens and a double bonded carbon, a double bonded carbon attached to two hydrogens, a double bonded carbon attached to two carbons, and a carbon attached to two hydrogens, a chlorine and a double bonded carbon.

Difficulties arise when a group identified as being in a molecule of interest cannot be found in the tables. To solve this problem the group may be fabricated by algebraic manipulation of existing groups.

An example of building the compound triethyl orthoacetate follows.

$$CH_3-C-(OCH_2CH_3)_3$$

Group	No. of Times Used
$C-(C)(H)_3$	4
$O-(C)_2$	3
$C-(C)(H_2)(O)$	3
$C-(C)(O)_3$	1 (Not available)

The orthoacetate group is not available; however, it can be created by algebraic combination of two other groups.

$$C-(C)(O)_3$$

$$\underset{\overset{|}{C}}{\overset{\overset{|}{C}}{O-C-O}} \;+\; \underset{\overset{|}{C}}{\overset{\overset{|}{C}}{O-C-O}} \;-\; \underset{\overset{|}{C}}{\overset{\overset{|}{C}}{C-C-O}} \;=\; \underset{\overset{|}{O}}{\overset{\overset{|}{O}}{C-C-O}}$$

Group	No. of Times Used
$C-(C)_2(O)_2$	2
$C-(C)_3(O)$	-1

Another type of substitution which can be used in CHETAH is molecular substitution. An example of this type is the estimation of the heat of formation of terephthalic acid from addition of two benzoic acids and subtraction of a benzene.

$$\Delta Hf° \text{ (Calculated)} = -158.5 \text{ kcal/mol}$$
$$\text{(Experimental)} = -161.4 \text{ kcal/mol}$$

The estimate is quite good, but is not necessarily typical of all such calculations. Much less confidence should be placed in data derived from this substitution technique.

Thermodynamic Calculations

The CHETAH data bank contains data for about 500 group contributions and about 400 chemical compounds. The data stored for the current program is heavily oriented toward organic materials. Data for chemical compounds is stored for 1) common, frequently used molecules, 2) molecules which are too small to be estimated by second order groups, 3) common decomposition products, and 4) molecules which can be used as building blocks to create other molecules. The results for a typical heat of reaction calculation are shown in Table II.

TABLE II.

THERMODYNAMIC DATA

Oxidation of ETOH
Temperature = 25°C
No Error Messages.

Reactant Compound(s)	AMT	MW	CP	DELHF	S	DELHC
Ethanol	1.00	46.07	15.43	-56.06	67.10	-305.4
/File Data Refs: 1,11,20						
Oxygen	3.00	32.00	6.99	0.0	49.00	0.0
/File Data Refs: 4,						
Product Compound(s)						
Carbon Dioxide	2.00	44.01	8.90	-94.05	51.10	0.0
/File Data Refs: 1,11,						
Water	3.00	18.01	7.99	-57.80	45.10	0.0
/File Data Refs: 1,11,						

Enthalpy of Reaction   -305.44 kcals
Entropy of Reaction   23.39 gibbs
Free Energy of Reaction   -312.41 kcals

Energy Hazard Evaluation

CHETAH makes an energy hazard evaluation by going through the following steps:

1. Estimate the enthalpy of formation of the compound from its structure.

2. Estimate the heat of combustion of the compound. It is assumed that oxygen and fluorine are the oxidizing elements. The fluorine in the molecule is used to convert the remaining elements to their fluoridized products, starting with the most electropositive elements. When all of the fluorine has been consumed, the remaining elements are carried to their fully oxidized products.

3. Estimate the maximum energy of decomposition. The program chooses the decomposition products from those flagged in the data bank as being possible decomposition products. It chooses a stoichiometric combination of products which will give the largest heat (the most exothermic).

4. Compute the difference between the heat of combustion and the maximum energy of decomposition.

5. Compute the oxygen balance with F_2 and O_2 as the oxidizing elements. The oxygen balance is a measure of the balance between the oxidizing and reducing components of the molecule and is an important characteristic of most compounds capable of exploding. Explosives are considered most sensitive when these components are present in approximately stoichiometric proportions. This balance is expressed as the percentage of oxygen required for complete conversion of the carbon and hydrogen in the molecule to carbon dioxide and water.

6. Use of pattern recognition to classify the compound as sensitive or insensitive.

CHETAH uses the above parameters along with experimental data to develop a classification scheme. The experimental data used were shock sensitivity data for about 218 compounds, of which 83 were known to be shock sensitive by standard drop weight or blasting cap tests. As an example of pattern recognition, if we wished to develop a classification scheme based on two variables, we would plot the experimental data versus the two variables. If we can define a region in which the yes values occur and in which the no values do not occur, then we have established a pattern. We could look at classification schemes requiring three, four, or more variables, although this becomes difficult, if not impossible, to plot. Therefore, we deal with the problem with a class of mathematical tools known collectively as pattern recognition.

The results for a typical energy release evaluation are shown in Table III.

The program gives a sensitivity rating of high, medium, or low for each of four correlations, and then it gives one overall rating for the compound or mixture.

Accuracy

With any estimation technique we need to know how good it is. For thermodynamic calculations, the heat of formation is usually accurate within ± 2 kcal/mol, the entropy within ± 2 cal/mol-K, and the heat capacity within ± 1 cal/mol-K. Error values exceed these in some cases especially for large molecules (those which require many groups to define) and for those having higher order effects. As with any group contribution method, occasionally a group needed to build a molecule will be missing. If another group is substituted, or a value is estimated from other sources, the error values should be increased. On energy hazard evaluations, CHETAH will make the correct classification 95% of the time. It is biased to err on the safe side four of the remaining five times.

Applications

Although CHETAH was designed primarily as a hazard evaluation tool, the thermodynamic data generated is of high enough quality to allow other applications.

TABLE III

Energy Release Appraisal   E.R.E. for TNT
Temperature = 25°C
Error MSG Code = 0

Reactant Compound(s)	AMT	M WT	HT FORMN	FORMULA
2,4,6-trinitrotoluene	1.00	227.13	10.99	$C_7H_5N_3O_6$

Product Compound(s)		
C	Carbon (Graphite)	5.25
N_2	Nitrogen	1.50
H_2O	Water	2.50
CO_2	Carbon Dioxide	1.75

Evaluation

Criterion No.	Value	Rating
1	-1.41	High
2	-2.17	High
3	-74.0	High
4	214.8	High

The energy release potential of this mix is *High*

A case in point is "dilution study". When using hazardous reactants such as benzoyl peroxide (BPO) or t-butyl hydroperoxide (TBHP) which are both labelled as hazardous by CHETAH, one can perform hazard evaluations at different concentrations in a stable (non-hazardous) solvent. This allows for selection of a safe concentration for initial scale-up studies in process development. The effect of dilution with xylene on the maximum enthalpy of decomposition of the above peroxides is shown below:

	ΔH_{max} (kcal/g)	
Molar Ratio	BPO	TBHP
1:1	-.63	-.68
1:4	-.55	-.55
1:15	-.50	-.49

It is apparent that ΔH_{max} decreases in magnitude and would eventually (at infinite dilution) approach that of pure xylene (-.46 kcal/g).

The CHETAH program assumes the gas phase but can also be used for evaluating condensed phases as well. When studying chemical reactions, if the assumption is made that the heat of vaporization of the reactants is equal to that of the products, the phase is irrelevant. This assumption is a reasonable one for fairly non-polar materials.

An example of where the CHETAH program was able to successfully be applied to the study of a condensed phase reaction is in the first step

$$\Delta H = -0.04 \text{ kcal}$$
$$\Delta S = 0.49 \text{ gibbs}$$
$$\Delta G = -0.23 \text{ kcal}$$

of Kondo's process for the manufacture of the synthetic pyrethroid intermediate ethyl 2,2-dimethyl-3-(dichlorovinyl)-cyclopropane-carboxylate (3). The chemistry is actually more complex than shown and involves three separate reactions, the first being a transesterification.

$$\Delta H = -0.04 \text{ kcal}$$
$$\Delta S = 0.49 \text{ gibbs}$$
$$\Delta G = -0.23 \text{ kcal}$$

The data for ΔH, ΔS, and ΔG of this reaction should be fairly good. The only group that was not available in the data tables was the orthoacetate group, $C(C)(O)_3$, but since that is present on both sides of the reaction it drops out of all calculations. The magnitude of ΔH for this reaction is quite small, so unlike hazard evaluations small steric corrections can have a major effect on the overall result and cannot be ignored. The second step,

$$\Delta H = 26.94 \text{ kcal}$$
$$\Delta S = 49.29 \text{ gibbs}$$
$$\Delta G = 8.55 \text{ kcal}$$

elimination of ethanol is more likely to be in error on an absolute basis because the orthoacetate group is present only on the left side of the equation.   However, the numbers agree with the expectation that this reaction would be endothermic and require heat input to drive off ethanol.

The third reaction, the oxy-Claisen rearrangement is very exothermic as most Claisen rearrangements are.   Once again we can place more confidence in the magnitude of the numbers because all the groups required are available in the data bank.

$$\Delta H = -33.70 \text{ kcal}$$
$$\Delta S = -5.34 \text{ gibbs}$$
$$\Delta G = -31.71 \text{ kcal}$$

The value of this type of analysis to the chemist and engineer is that in a matter of minutes he can have a significant amount of data that can be used in analyzing a process.   Besides evaluating a process on expected yields and raw materials costs, it is possible to predict some energy and capital costs for heating and/or cooling. Also, in planning to test the reaction in the laboratory, the chemist should be prepared to handle the potential problem from the release of energy in the last step and should not spend a significant time looking for the second step product as it is expected to react as rapidly as it is formed.   This material was not detected during actual experiments.

It is important for the scientist who is extending the utility of CHETAH in this manner to keep in mind earlier cautions that the data obtained is a computer estimate only and is not meant to replace appropriate physical tests.

Overview of New CHETAH Program

CHETAH, through the efforts of the committee is a dynamic program that will undergo many changes for the next published version.   A brief overview of the upcoming version is as follows. The number of elements will be expanded from 22 to 74.   The new version includes temperature dependent data for reference state elements, including phase change information.   The Wilcox-Bromley method, which uses ionic group contributions to estimate data for inorganic compounds, has been added; this will greatly expand the capabilities of the program for estimating data for inorganic species.   Data for at least 100 new groups and about 100 whole molecules has been added to the data bank.   Currently about a dozen new correlation schemes for determining the potential hazard are being evaluated using pattern recognition techniques.

Summary

In summary, we would like to emphasize the two current capabilities of the CHETAH program. The first is the estimation of gas phase thermodynamic data over the range of 25-1200°C, using Benson's second order group contribution method. The second capability of CHETAH is a hazard appraisal. This uses thermodynamic data to determine the maximum energy available in the compound or mixture and then rates the potential hazard using a rating scheme based on experimental shock sensitivity data. Although the results compare well with experimental data, once again we reiterate that this should not be used as the sole test for hazard evaluation, but merely as a guide for further testing.

Literature Cited

1. Seaton, W. H., Freedman, E., and Treweek, D. N., "CHETAH - The ASTM Chemical Thermodynamic and Energy Release Evaluation Program", ASTM Data Series Publication DS-51, 1974.

2. Benson, S. W., "The Thermochemical Kinetics", Wiley and Son, New York, 1968.

3. Kondo, K., U. S. Patent 4,214,097, 1980.

RECEIVED November 3, 1984

Kinetic and Reactor Modeling

Hazard Evaluation and Scale-up of a Complex Reaction

ASHOK CHAKRABARTI, EDWIN C. STEINER, CRAIG L. WERLING, and
MAS YOSHIMINE

The Dow Chemical Company, Midland, MI 48640

Two separate models based on Dow Advanced Continuous Simulation Language (DACSL) were used in these studies. The first model used laboratory data and parameter estimation to determine the Arrhenius constants for two desired and eight undesired reactions in a process. The second model used the Arrhenius constants, heats of reaction, different physical properties, and reactor parameters (volume, heat transfer properties, jacket control parameters, jacket inlet temperature) to simulate the effect of reaction conditions (concentration, set temperature, addition rate) on the temperature of the reaction mixture, pressure and gas flow rates in the reactor, yield, and assay of the product. The program has been successfully used in two scale-ups where the optimum safe operating conditions, effect of various possible failures, and control of possible abnormal conditions were evaluated.

Hazard analysis for a process normally involves a battery of tests including Differential Scanning Calorimetry (DSC) and Accelerating Rate Calorimetry (ARC). Optimization and scale-up also require extensive experimental work in reactors of different sizes with different temperatures, compositions, etc. Time commitment and difficulty in interpreting the results will depend on the complexity of the process. Under most circumstances, a modeling approach is cost-effective (1). The importance of modeling will actually increase with the increasing complexity of the process.

The reactions to be considered here are shown in Figure 1. Of the ten or so simultaneous reactions, only two are desired. Although no chemical names will be used to protect proprietary information, it is felt that the usefulness and the capabilities of the model can be explained properly with letters as names.

The usefulness of DSC and ARC data in this case was limited for several reasons: 1) It is next to impossible to reproduce the

0097-6156/85/0274-0091$06.00/0

MAJOR REACTIONS

                                          Heat of Reaction
                                             kcal/mole

*   A + B $\xrightarrow[K1R]{K1F}$ C + D   -4

        A $\xrightarrow{K2}$ C + GAS   -25

        D $\xrightarrow{K3}$ B + GAS   -25

*   RCT + D $\xrightarrow{K4}$ PDT + B   -65

MINOR REACTIONS

        RCT + D $\xrightarrow{K5}$ E + B   -15

** E + B $\xrightarrow{K6}$ F   -10

** PDT + G $\xrightarrow{K7}$ H   -10

** RCT + D $\xrightarrow{K8}$ I + B   -60

        RCT $\xrightarrow{K9}$ G + Tar

        PDT $\xrightarrow{K10}$ G + Tar

            Three other equilibria with known
                  equilibrium constants

Figure 1.   Reactions in the process for which individual rate
constants were determined.   *Represents two desired reactions.
**Reactions producing major impurities totaling up to 10% of
product.

proper product mix under conditions of testing; 2) The reaction
mixture is heterogeneous; 3) Two undesired reactions (2 and 3) are
significant well below the desired reaction temperature and are
exothermic; 4) The actual hazard of the process is compounded by
the similarity of the rates of reaction 1F and 4, where the product
(D) from a reaction producing very little heat (reaction 1F) is
needed for a major exothermic reaction.

In this paper we will discuss the application of a general
batch reactor model that considers the reaction kinetics, heats of
reaction, heat transfer properties of the reactor, physical
properties of the reactants and the products, to predict: 1) The
concentration profile of the products, thus enabling process
optimization; 2) Temperature profile during the reaction, which
provides a way to avoid conditions that lead to a thermal runaway;
3) Temperature profile of the jacket fluid while maintaining a
preset reactor temperature; 4) Total pressure in the reactor, gas
flow rates and partial pressure of different components. The model
would also allow continuous addition of materials of different
composition at different rates of addition.

The model is divided into two parts. The first one is called
the KINETIC model and is used only for determining the rate
constants of the different reactions. The second model, called the
REACTOR model, is the work-horse and will be the focal point of the
present discussion.

In order to provide a better understanding of the modeling
language and its simplicity, the actual variable names and
expressions from the programs have been used throughout this paper.
Since the variables are somewhat difficult to follow, a directory of
these names is supplied.

Data

Isothermal laboratory reactors with 200 gm of reactants were used in
these studies. Instead of continuous addition (as envisioned in the
production reactor), component A was added in three shots over a
period of 20 minutes. Concentration of various components were
followed by calibrated GC. The amount of gas formed was measured by
a wet test meter. Multiple runs at each of three different
temperatures were used to generate the necessary kinetic data. For
convenience, concentrations were expressed in mol/kg.

For the reactor model, individual specific heats and vapor
pressures were either measured or obtained from the literature. All
heats of vaporization were obtained from the literature. The heats
of reaction were either measured, obtained from literature, or
estimated by conventional means.

Nature of the Model

Both models are based on Dow Advanced Continuous Simulation Language
(DACSL) (2). Though DACSL itself is proprietary in nature, a
significant portion of it is based on commerically available
packages. DACSL is designed for analysis of the dynamics of
physical systems by people with limited computer background. A
short discussion on the most important features of DACSL would help
in understanding the present models.

Ordinary differential equations are solved by an operator INTEG, where the integration is started by the command START. Six algorithms are available for this purpose.

Parameter estimation, that is the extraction of individual rate constants, was accomplished by a command ESTIMATE. Here the rate constants are varied until the best fit between the observed and calculated values are obtained. With the rate constants in this system covering five orders of magnitude in range, and somewhat limited amount of concentration data, it was often necessary to change the rate constants after the ESTIMATE procedure. This is done interactively through repeated use of SET, START, and PLOT commands. One such plot is shown in Figure 2. The lines in Figure 2 represent the calculated values, whereas the points represent the observed values. Arrhenius parameters for individual reactions are then calculated from the rate constants at three temperatures. These constants are then used in the REACTOR model.

The rate of each reaction (R1F, R2, etc.) is calculated from the rate constants and the concentrations. The rate of change of concentration of each species is then calculated by combining the individual rates of reactions which involve this species. For Species B: $XB = -R1F + R1R + R3 + R4 + R5 - R6 + R8$. Concentration for species B is then calculated by integrating XB from an initial concentration of IB: $B = INTEG(XB,IB)$. Similar rate and integration statements will appear for each species in the reaction.

The reaction mixture, for most of the reaction period, forms a two phase system. A few observations/assumptions allowed the treatment of the conditions as homogeneous. That is, the concentrations were calculated on the basis of the total mixture, even though one component may reside mostly in one phase or the other. These assumptions are: 1) The reaction is not limited by mass transfer. With the available stirring this was found to be true. 2) The ratio of the weight of the two phases remains constant. 3) The partition coefficients remain constant throughout the reaction. Except for a small part of the total reaction time, conditions (2) and (3) were found to be true.

A typical batch reactor is shown in Figure 3. The various features of such a reactor introduced as variables are also shown here. The jacket has a heat transfer coefficient of U and an area of AJ. The temperature of the jacket inlet is TJIN, flow rate of jacket fluid is FLOWJN. The capacity of the jacket and the heat capacity of the jacket fluid are also considered. The gain and reset parameter for the valve controlling TJIN are GAINJ and TAUIJ. TSET is the set temperature of the reactor. TJ is the temperature of the fluid leaving the jacket. TRX is the temperature of the reaction mixture ²nd is assumed to be constant throughout the reactor. The temperature of the reaction is calculated from the enthalpy (NTHLP) and heat capacity of the reaction mixture. Rate of change of enthalpy (XNTHLP) is calculated by combining the terms for rate of heat generation by reaction (XQRX), heat lost through the jacket (XQTR), sensible heat (XQSENS) via addition of material (ADRATN), heat from the stirrer (XQSTIR), vaporization (XQVAP), material returned from the condenser (XQRFLX), and fraction of vapor returned to the condenser (1-RFLXN).

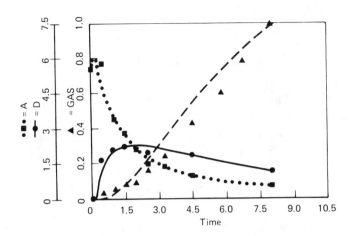

Figure 2. Comparison of calculated and observed concentrations.

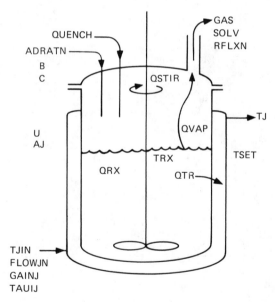

Figure 3. A typical batch reactor with the major variables considered in the model.

$$XNTHLP = XQRX - XQTR + XQSENS + XQSTIR - XQVAP + XQRFLX(1-RFLXN)$$

Integration of the enthalpy term from a starting value (INTHLP, calculated from initial conditions) yields the enthalpy at different times.

$$NTHLP = INTEG(XNTHLP, INTHLP)$$

The physical properties needed for these (and other) calculations are entered as function of temperature. Vapor pressure is calculated by Antoine equation, e.g.,

$$VPB = 10**(A1B - B1B/(TRX + C1B))$$

Heat of vaporization is expressed as: HVB = A2B + TRX*(B2B + C2B*TRX)

Specific heat of B is calculated from: CPB = A3B + B3B*TRX

Jacket inlet temperatures are calculated by conventional proportional and integral fucntions with an upper and lower bound.

$$
\begin{aligned}
ERROR &= TSET - TRX \\
ERINT &= INTEG(ERROR, 0.0) \\
TJIN &= TSET + GAINJ*ERROR + ERINT/TAUIJ \\
TJIN &= BOUND(TJMIN, TJMAX, TJIN)
\end{aligned}
$$

Pressure is calculated for isothermal flow of gas through a round horizontal pipe (3). Flow of gas in mol/hr (FLVAP) through a vent can be expressed as

$$FLVAP = \sqrt{\frac{2.93E5*(PRESS**2-1)*(D**5)}{AVMWT*F*L*(TRX+273)}}$$

Rate of accumulation of gas in the head-space (XPGASH) is calculated by subtracting the rate of loss of GAS through the vent from the rate of generation of GAS (XMGAS). It is assumed that the head-space is always saturated with the vapors of reactants and products (PSOLV). Total pressure in the reactor is the sum of PSOLV and the partial pressure of gas left in the reactor (PGASH)

$$
\begin{aligned}
XMGASH &= XMGAS - FLVAP*PGASH/PRESS \\
MGASH &= INTEG(XMGASH, IMGASH) \\
PGASH &= MGASH* RG *(TRX+273)/VFREE \\
PRESS &= PSOLV + PGASH
\end{aligned}
$$

Operation of the Reactor Program

Initial loading of the reactor and the variables discussed in this section are the only quantitites changed to simulate the operation of a particular reactor. All other parameters in the REACTOR model are typical of either the reaction or the reactor. The initial reactor contents was the same in all cases.

Time Segments. There are six time segments within which several
variables can be changed to simulate an actual reaction condition.
The length of each segment is changed by a variable STAGE. The
duration of a stage is the difference between the numerical values
of two successive values of STAGE. The total number of available
stages can be increased with some minor changes in the program. The
variables that can be changed within each stage are discussed below.
Though more than one of the variables can be changed in each STAGE,
a STAGE is needed to make any change.

Continuous Addition. The rate of addition of a component or mixture
of components is expressed by the variable ADRATN.

Composition of Added Material. This composition can be changed by
variables FCA, FCB, FCC, FCD (all in mol/kg). If the added material
has no B in it then FCB = 0, and so on. Quench is pure C, such that
FCC = 60, FCA = FCB = FCD = 0 will be used in the STAGE representing
addition of quench material.

Set Temperature of the Reactor. Changed by changing TSET.

Parameters for Jacket Control. Gain = GAINJ, Reset = 1/TAUIJ.

Reflux Conditions in the Condenser. This is somewhat qualitative in
that heat transfer in the condenser is not considered, but only the
fraction of vapor returned to the reactor is expressed as (1-RFLXN).

Flowrate of Fluid Through the Jacket. FLOWJN is used to simulate
the loss of jacket coolant or simply the size of the jacket.

Simulation of Reaction Conditions

Given the parameters for a reactor, the variables discussed above
are changed to represent the conditions of the process. An
optimization process, whose goal is to achieve the best assay for
the product as fast as possible in a safe manner, actually considers
the hazard and optimization process simultaneously. No attempt was
made to find the optimum by computational means, instead a "manual"
approach was taken. This was done interactively by observing the
effect of a change made and by following the trend.
    The condition of the reaction is such that we would like to add
a large excess of A in one shot and operate the reactor between 80
and 90°C. The main hindrance to this is the heat transfer
capabilities of the reactor and a possible rise in pressure in the
reactor.
    We will next discuss a few typical uses of the model for
optimization and hazard evaluation. The results shown are all
predicted values, the points shown in these figures are only for
identifying different curves.

Optimization. An ideal addition will require gradually decreasing
addition rate of A followed by a gradual increase after a minimum.
This is simulated by a stepped addition shown in Figure 4. The
numerical values of the variables for this case are given in
Table I.

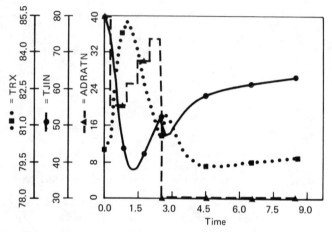

Figure 4. Temperature profiles of the reaction mixture and the
jacket fluid at the inlet for a stepped addition of the feed.

TABLE I. Numerical Values of Variables Used to Generate
          the Simulation in Figure 4

STAGE	0.25	1.0	1.5	2.0	2.5	15
ADRATN	40	20	25	30	35	0
FCA	15	15	15	15	15	15
FCC	20	20	20	20	20	20
TSET	80	80	80	80	80	80
FLOWJN	2000	2000	2000	2000	2000	2000
DIN			0.5			

The first STAGE is 0.25 hours. That is, for this time a mixture
containing 15 mol/kg of A (FCA) and 20 ml/kg of B (FCB) was added at
a rate of 40 kg/hr (ADRATN); the set temperature of the reactor for
this time period was 80°C (TSET); flow of fluid through the jacket
was 2000 kg/hr (FLOWJN); the diameter of the vent pipe cannot be
changed over different STAGEs and was 0.5 in (DIN). These variables
have the values listed in the second column for the next (1.0-0.25)
or 0.75 hours. For each of the next half-hour segments, the
addition rate increased by 5 kg/hr up to 35 kg/hr, stopping
altogether after 2.5 hours from the start of the reaction. The
calculated maximum temperature of the reaction mixture was 85°C.
The manimum jacket inlet temperature was 38°C, which is above the
available minimum fluid temperature of 30°C. This reaction is
considered to be under control. Figure 5 shows the corresponding
concentration profile for the desired product (ASPDT), and species A
and D. The concentration of the product in this case is plotted as
assay of the product in the organic phase. Since the peak
concentration of the product can be predicted, the model helps in
determining the desired time for the cool-down of the reactor.

A Runaway. The conditions for the reaction simulated in Figure 6
are similar to that in Figure 4, except for an addition of 7% more
of A than in the earlier case. A runaway is depicted by an increase
in the temperature of the reactants (112°C) and bottoming out of the
jacket inlet temperature at 30°C. It is important to note that the
runaway starts half an hour into the reduced addition rate. This
observation led to the conclusion that a previously (pre-modeling)
suggested operation of controlled addition by monitoring temperature
change is not feasible.

Temperature Stepping. Figure 7 shows a simulation where the set
temperature for the reactor was changed. The addition was
accomplished at two rates of 22 and 24 kg/hr for 2.5 hours each.
The set temperature (TSET) for the first 2.5 hours was 80°C and an
excursion of 3°C was observed in the simulation of this time period.
The set temperature was changed twice more to 85°C for the next 2.5
hours and 90°C for the rest of the reaction. The calculated
deviation from TSET in those two steps was caused by deliberate use
of poor gain and reset parameters for the jacket control. This
reaction is acceptable from the viewpoint of safety. Though only
one or two plots are shown for each run, in actuality several other

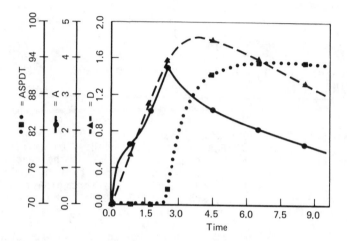

Figure 5.   Concentration of three components during the
reaction presented in Figure 4 and Table I.

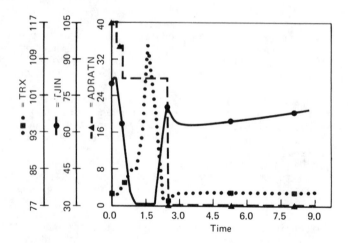

Figure 6.   A thermal runaway caused by high addition rate.

parameters are followed, the most notable of which is assay of the product.

Instantaneous Addition. Two scenarios need to be considered from a safety standpoint: 1) Failure of the control mechanism of the addition of A, where a limiting orifice was not used. 2) An operator forgetting to start the stirrer before starting the addition. This becomes very important because the reaction mixture will form two separate phases. (This is different from the pseudo-homogeneous treatment of the kinetics).

This situation is simulated by assuming that all of A is added in 15 minutes. Figure 8 shows the resultant temperature for a total of 75 kg of A. A sensible cooling followed by a runaway is observed. A boiling of the reaction mixture is expected as a nearly constant temperature of 118°C for approximately 10 minutes is seen in the plot of TRX.

Loss of Jacket Fluid. A possible equipment failure involves partial or total loss of flow in the jacket. This is simulated by varying FLOWJN. In the simulation presented in Figure 9, the jacket failure was detected at 2 hours (FLOWJN=0 after the first STAGE of 2 hr). It is assumed that it took the operator 15 minutes to turn off the proper valves and to add the quench material C. A total of 75 kg of C was added over a 15 minute period (300 kg/hr). Addition of C helps in three different ways: 1) Sensible cooling; 2) Increased reverse reaction R1R; and 3) Reduction of overall concentrations. In this case it is seen that 75 kg of C will not be enough. The temperature rise after the addition of quench is due to reaction still under progress. Several such plots with quench added after different lengths of time into the reaction will be needed to decide on the size of the quench tank.

Different Vent Sizes. The effect of a closed vent was simulated by using 0.01 inch as the diameter of the vent pipe. The calculated pressure and temperature are shown in Figure 10. For the particular reactor considered, a pressure in excess of 27 atm is expected. Trials with other diameters for vent pipe showed that a diameter of 0.1 inch will be adequate under normal operating conditions.

Summary

Through the use of a model for a batch reactor for a particularly complex reaction, we have demonstrated the value of modeling in optimization of process conditions and in evaluation of possible hazards. For a very complex system like the present one, it is most probably easier and more cost effective to do the modeling than to run the experiments needed for proper analysis. To save laboratory data acquisition time, it is always better to plan an experimental strategy based on the anticipated need in advance. This model has been successfully used in two scale-ups. Data from these scale-ups have been used to refine the model. These refinements included a better understanding of the chemistry of the process. Plots similar to the ones presented in Figures 6-10 were used in the Reactive Chemicals Review of the present process.

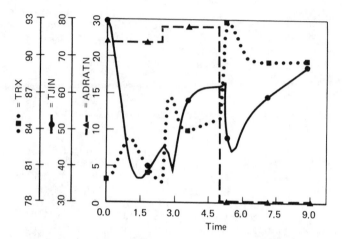

Figure 7. Effect of changing set temperature on the temperture of the reaction mixture and the jacket fluid at the inlet.

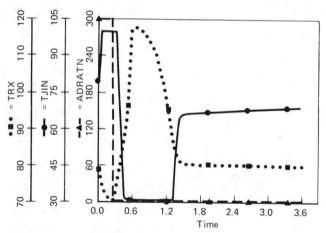

Figure 8. Thermal runaway caused by shot addition.

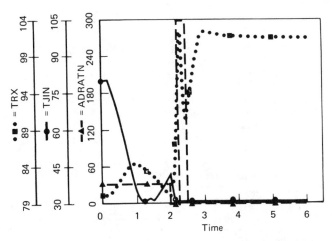

Figure 9.   Temperature profiles for loss of jacket fluid after two hours and addition of quench.

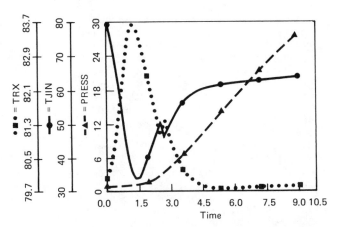

Figure 10.   Pressure rise due to a closed vent for an otherwise normal reaction.

With the availability of various computer packages, it has become very easy for chemists to perform in-depth analysis of chemical systems. The present models can be adapted to other systems via proper rate expressions, physical properties and heats of reaction. Work is under progress to expand this model to a series of CSTRs.

Notation

ADRATN	Rate of addition of reactants or quench.
AJ	Area of the jacket.
ASPDT	Assay (wt%) of the product (PDT) in the organic phase of the reaction mixture.
AVMWT	Average molecular weight of the gases and vapors in the head space of the reactor.
A1B	Antoine constant for species B.
A2B	Constant for heat of vaporization for species B.
A3B	Constant for specific heat of species B.
BOUND	DACSL operator keeping jacket temperature between specified upper and lower limits.
B1B	Antoine constant for species B.
B2B	Constant for heat of vaporization of species B.
B3B	Constant for specific heat of species B.
CPB	Specific heat of species B.
C1B	Antoine constant for species B.
C2B	Constant for heat of vaporization for species B.
D	Diameter of the vent pipe.
DACSL	Dow Advanced Continuous Simulation Language.
DIN	Diameter of vent pipe, inches.
ERROR	Difference of the actual temperature of the reaction mixture from the set temperature of the reactor.
ESTIMATE	DACSL command for evaluating constants by curve fitting.
F	Friction factor of the vent pipe.
FCA	Moles per Kg of species A in the feed.
FCB	Moles per Kg of species B in the feed.
FCC	Moles per Kg of species C in the feed.
FCD	Moles per Kg of species D in the feed.
FLOWJN	Flowrate of fluid through the jacket.
FLVAP	Total flow of gases and vapors through the vent pipe.
GAINJ	Gain parameter controlling the temperature of the jacket fluid.
GAS	Non-condensible reaction product from reactions 2 and 3.
HVB	Heat of vaporization of species B.
IB	Initital concentration of species B.
IMGASH	Initial moles of gas in the head space.
INTEG	DACSL operator for integration.
INTHLP	Enthalpy of the reaction mixture in the reactor at the start of the reaction.
L	Length of a horizontal vent pipe in feet.
MGASH	Moles of GAS in the head space.
NTHLP	Enthalpy of the reaction mixture.
PDT	Desired final product.
PGASH	Pressure of GAS in the head space.
PLOT	DACSL command for plotting experimental and calculated values of a variable.

PRESS	Pressure in the head space, atm.
PSOLV	Total of the vapor pressures of reactants and products (except GAS).
QUENCH	Material C at ambient temperature to be added for control of thermal runaways.
RCT	Reactant.
RFLXN	Fraction of vapor not condensed by the condenser and escaping the reactor.
RG	Gas constant, lit.atm/mol.K.
R1F	Rate of the forward reaction involving A and B.
R2	Rate of reaction 2 producing C and GAS.
SET	DACSL command for assigning new values for constants.
SOLV	Vaporized reactants and products.
STAGE	Different time segments of the total reaction time. STAGEs need to be assigned for changing several reactor parameters and conditions.
START	DACSL command for starting integration.
TAUIJ	Inverse of reset parameter for controlling the temperature of the jacket fluid.
TJ	Actual temperature of the jacket fluid at the exit.
TJIN	Temperature of the jacket fluid at the inlet.
TJMAX	Maximum attainable temperature of the jacket fluid.
TJMIN	Minimum attainable temperature of the jacket fluid.
TRX	Temperature of the reaction mixture.
TSET	Set temperature of the reactor.
U	Heat transfer coefficient for the reactor.
VFREE	Volume of head space in the reactor.
VPB	Vapor pressure of component B.
X...	Time derivative of ...
XB	Rate of change of concentration of B.
XMGAS	Rate of generation of GAS, mol/hr.
XMGASH	Rate of accumulation of GAS in the head space.
XNTHLP	Rate of change of total enthalpy.
XQRFLX	Rate of change of enthalpy due to refluxed material.
XQRX	Rate of generation of heat from reaction.
XQXENS	Rate of change of enthalpy due to feed.
XQSTIR	Rate of heat generation by the stirrer.
XQTR	Rate of heat loss through the jacket.
XQVAP	Rate of loss of heat due to vaporization.

Literature Cited

1. Silverstein, J. L.; Wood, B. H.; Leshaw, S. A. Loss Prevention, 1981, 14, 78.
2. Agin, G. L.; Blau, G. E. AIChE Symp. Ser., 1982, 78(214), 108.
3. Perry, R. H.; Chilton, C. H. "Chemical Engineers' Handbook", 5th ed., McGraw Hill: New York, 1973; Equation 5.61.

RECEIVED November 13, 1984

The Nitration of 5-Chloro-1,3-dimethyl-1*H*-pyrazole
Risk Assessment Before Pilot Plant Scale-up

JAMES R. ZELLER

Pharmaceutical Research, Parke-Davis Division, Warner-Lambert Company, Holland, MI 49423

Reaction conditions which allowed for the large scale
nitration of 5-Chloro-1,3-dimethyl-1H-pyrazole were
developed which minimized the hazards generally
associated with nitration reactions. Dilution with
sulfuric acid decreased the risk of thermal instability.
Using ordinary laboratory equipment, the experimental
heat of reaction was determined to be -12.5 Kcal/mole.
Likewise, the adiabatic temperature rise was found to be
about 20°C. An exotherm was found to initiate at 100°C.
The thermal stability and shock sensitivity of the
product, 5-Chloro-1,3-dimethyl-4-nitro-1H-pyrazole, was
investigated using simple tests.

The introduction of a reaction to the pilot plant often proceeds
without determining the chemical hazards involved with the scale up
process. The reasons for this vary. In cases where a hazards
evaluation laboratory is not available, it is the responsibility of
the development chemist to assure the safety of the reaction. The
development chemist may not be familiar with hazard evaluation
techniques, and the instrumentation used to evaluate a reaction for
safety may not be readily available.

In our laboratory, we had the assignment of developing a
process to produce 10 Kg of a potential drug which required the
scale up of the nitration reaction of 5-Chloro-1,3-dimethyl-1H-
pyrazole (CDMP, equation 1). The nitration of CDMP was originally
carried out on small scale by medicinal research chemists. Using
those same conditions during scale up to the 20 mole level, a large
exotherm was observed leading to much foaming and loss of product.
Our goal was two-fold: (1) to develop nitration conditions which
would be safe upon scale up; and (2) to test this reaction for
safety (1,2). The steps taken to accomplish these goals will be
described.

0097–6156/85/0274–0107$06.00/0

We were hesitant to work with this reaction until we were confident that the nitration system did not possess an appreciable detonation potential. Although there are thermochemical computer programs available which can calculate the decomposition process which yields the maximum energy release of a system ($\underline{3}$), we did not have such a program available. We therefore estimated the potential detonation energy of the nitration system as described by Chester Grelecki of the Hazards Research Corporation ($\underline{4}$). Briefly, this technique consisted of balancing the chemical equation of the most plausible decomposition reaction, determining the heat of the reaction, and calculating the TNT equivalence. Since the oxygen fuel balance for the reactants was calculated to be unity (based on the ratio: Oxygen/(2Carbon + 1/2 Hydrogen)), we calculated the heat of decomposition for both the oxygen rich case and the oxygen poor case. In the oxygen poor case, the major carbon containing product is CO, and based on the stoichiometry of the reactants, the equation for the decomposition of the nitration reaction may be written as follows:

$$C_5H_7ClN_2 + 2\ HNO_3 + 2.3\ H_2SO_4 + 0.7\ H_2O \longrightarrow$$
$$7\ H_2O + 5\ CO + 2\ N_2 + 2.3\ SO_2 + HCl$$

The ΔH (decomposition) was calculated from the known ΔH_f of the reactants and products, found in a Physical Chemistry text ($\underline{5}$). The ΔH_f of DMCP was estimated to be 38.8 kcal/mole by the CHETAH program ($\underline{6}$). The ΔH (decomposition) of TNT in an oxygen poor system is -650 cal/g ($\underline{4}$), and the TNT equivalence was calculated as follows:

$$\Delta H(\text{decomposition}) = \Delta H_f(\text{products}) - \Delta H_f(\text{reactants})$$
$$= -721\ \text{kcal/mole} - (-534\ \text{kcal/mole})$$
$$= -187\ \text{kcal/mole}$$

Total weight of a 1 mole run = 494 g; $-187/494 = -0.378$ kcal/g
TNT equivalent   =   378/650 or 58.1%

For the oxygen rich case, the major carbon containing product is CO_2, and the heat of decomposition of TNT is taken to be -1100 cal/g ($\underline{4}$). The equation for this case may be written as follows:

$$C_5H_7ClN_2 + 2\ HNO_3 + 2.3\ H_2SO_4 + .7\ H_2O \longrightarrow$$
$$5\ H_2O + 4\ CO_2 + 2\ N_2 + SO_2 + 1.3\ H_2S + HCl + CO$$

$$\Delta H(\text{decomposition}) = \Delta H_f(\text{products}) - \Delta H_f(\text{reactants})$$
$$= -790\ \text{kcal/mole} - (-534\ \text{kcal/mole})$$
$$= -256\ \text{kcal/mole or} -0.518\ \text{kcal/g}$$

TNT equivalent   =   518/1100 = 47%

Based on these results, we assumed that the system possessed some thermal instability, and our strategy was to lower the heat of decomposition of the nitration system. Dilution of unstable systems tend to increase their stability. Calculation of the heat of decomposition, after dilution with 15 molar equivalents of sulfuric acid (versus 2.3 molar equivalents in the original procedure) shows that the potential for explosive behavior is greatly diminished. This is reasonable since sulfuric acid decomposes endothermically and is not an oxidizing agent.

$$C_5H_7ClN_2 + 2\ HNO_3 + 15\ H_2SO_4 + 5\ H_2O \longrightarrow$$
$$5\ CO_2 + 24\ H_2O + 15\ SO_2 + HCl + 2\ N_2 + 3.5\ O_2$$

$$\triangle H(\text{decomposition}) = \triangle H_f(\text{products}) - \triangle H_f(\text{reactants})$$
$$= -2943\ \text{Kcal/mole} - (-3294\ \text{Kcal/mole})$$
$$= +351\ \text{Kcal/mole}$$

Total weight of a 1 mole run = 1835 g

H(decomposition) = + 191 cal/g reaction mixture

The thermal stability of the nitrated product, 5-Chloro-1,3-dimethyl-4-nitro-1H-pyrazole (CNP), was tested in the laboratory by very simple tests. The product pyrazole (0.1 gm) was placed on a hot plate preheated to 300°C. The CNP melted and decomposed giving off a white smoke. Such results can be taken as negative. We investigated the shock sensitivity of the product by placing a few crystals on a steel plate and hitting the crystals with a carpenter's hammer (1). Again the results were negative. We placed a 6 inch strip of the compound on a watch glass and ignited the compound with a propane torch. The material burned very reluctantly and self extinguished after removing the torch. There were no indications of explosive tendencies. Finally, we heated a 20 g sample of CNP to 250°C. The only area of thermal activity observed was near 75°C, the melting point of the compound. In this series of tests, only positive results would have been conclusive, while negative results did not prove that the material was safe to handle. It cannot be over emphasized that if there were any indications that the product CNP was found to be thermally unstable or that the nitration system could not be designed to be safe, all work would have been stopped on the project until more sophisticated analysis (ARC, DSC, card gap test, etc.) indicated that it was safe to continue our study.

At this point, laboratory experiments were performed to tune the nitration reaction for yield and safety. We found that not only did excess sulfuric acid lower the detonation potential of the reaction, but it was also beneficial for other reasons. Sulfuric acid increased the rate of the reaction by increasing the concentration of nitronium ions; thus allowing the reaction to occur at a lower temperature (30°C with 15 molar equivalents, versus 90°C with 2.3 molar equivalents) where it was less likely to exhibit instability. Sulfuric acid also tied up the water in the system, which is known to deactivate nitration processes (7), and allowed for the use of 70% nitric acid, which was easier and safer to handle than 90 or 95% nitric acid. It was found that when the basic CDMP was mixed with sulfuric acid, a large amount of heat of neutralization was evolved. The mixing of 70% nitric acid with 96% sulfuric acid also evolved heat. Premixing the pyrazole with one half of the sulfuric acid, and premixing the nitric acid with the remainder of the sulfuric acid, removed these sources of heat from the nitration system.

These observations were incorporated into the following nitration procedure. 5-Chloro-1,3-dimethyl-1H-pyrazole (133 g, 1 mole) was dissolved with cooling in 96% sulfuric acid (686 g, 7 mole). This solution was added, maintaining a temperature of 30°C, to a solution of 70% nitric acid (177 g, 2 mole) in 96% sulfuric acid (784 g, 8 moles). The reaction was stirred for 20 minutes, poured onto ice water, and the product was collected by filtration. Only slight cooling was necessary for maintaining the 30°C reaction temperature, and there was no evidence, such as discoloration or bubbling, to suggest that decomposition had occurred. It was this system which we decided to test further for thermal instability.

Initial experiments conducted to test the thermal stability of the nitration reaction were designed to determine if exotherms occurred at elevated temperatures. Behind a barricade, we remotely heated about 6 grams of the nitration reaction mixture in a silicone oil bath to 260°C while recording the temperature of the bath, the temperature of the sample, and noting any observations. (Table I). We observed an exotherm initially (from the heat of the nitration reaction) and another exotherm at about 100°C. This initial experiment was repeated using three different CDMP charge ratios (33, 50, and 66% mole ratios) to represent the reaction during the addition process. The thermal behavior did not differ significantly from the original case, where the full molar equivalent of the CDMP was present.

Table I. Open Test Tube Thermal Stability Test

Time (min.)	Temp. Bath (°C)	Temp. Sample (°C)	T	Observations
0	24	24	0	Added CDMP
5	25	32	+7	
10	26	33	+7	
15	26	29	+3	
20	27	27	0	Initiated heating
30	73	53	-20	
40	128	110	-18	
45	148	135	-12	Bubbling noticed
50	163	157	-6	More bubbling
55	175	170	-5	Brown Fumes
60	213	184	-29	Fumes stop
65	233	218	-15	Heating stopped

Knowing that exothermic activity did exist, it was now necessary to further define the temperature of initiation of the exotherm using a system which approximated the near adiabatic conditions found in a jacketed 50 gallon reactor. We constructed a calorimeter consisting of a 250 ml 3-neck round bottom flask immersed in a silicone fluid bath, all contained in a dewar flask (1, Figure 1). The temperature of the bath could be increased by use of a Nickel-Chrome wire heating element connected to a variable power supply. Both the temperature of the bath and the sample were recorded on a dual pen chart recorder. Both the bath and the sample were stirred. A 150 ml sample of the nitration mixture was heated at a rate of 3°C/min. and the first exotherm was noted to begin at 100°C. This exotherm peaked at 183°C at which point the temperature of the sample was 12°C above the temperature of the bath. Another exotherm was observed starting at 220°C and peaked at 270°C where the sample temperature was 14°C above the temperature of the bath (Figure 2). No thermodynamic data could be obtained from this experiment since gases were allowed to escape and the exotherms were probably moderated by this slow endothermic vaporization. Nevertheless, it was very important to know that if the temperature of the nitration was allowed to approach 100°C, we could expect exothermic behavior. We therefore had to assure that the reaction temperature could not reach 100°C.

The temperature rise of an exothermic reaction is dependent on three factors: the heat of the reaction, the heat capacity of the system, and the heat loss of the system. The temperature rise of a reaction in a system with no heat loss, the adiabatic temperature rise ($\triangle$T), is dependent on the heat of the reaction and the heat capacity of the system, and independent of scale. To determine the adiabatic temperature rise of this system, the CDMP/sulfuric acid solution, prewarmed to 30°C, was added all at once to a dewar flask containing the nitric acid/sulfuric acid solution which was also prewarmed to 30°C. We observed a temperature rise of 17°C over a period of 4 minutes, with a temperature drop of 1.5°C over the next 4 minutes (Figure 3). We therefore estimated the $\triangle$T to be about 18.5°C. Since this temperature rise was, in theory, independent of scale, we could predict that the large scale nitration reaction would not rise to a temperature of exothermic activity. Based on these results, we considered this revised nitration procedure to be safe upon scale up to the pilot plant.

At this point, we returned to the original nitration procedure (with 2.3 molar equivalents of sulfuric acid) to try to determine why the nitration was not safe upon scale up. The $\triangle$T for this original procedure could be calculated from the heat capacity (Cp) of the system and the heat of the reaction ($\triangle$H) by the equation $\triangle$H = Cp x $\triangle$T. The heat capacity of nitric acid/sulfuric acid/water systems are available (8) and found to be 678 cal/mole °C (assuming that the heat capacity of the CNP/CDMP component was negligible). Using the experimentally derived temperature rise, the heat of the nitration reaction was estimated to be:

$$\triangle H = (18.5°C)(0.678 \text{ Kcal/mole } °C)$$
$$= 12.5 \text{ Kcal/mole}$$

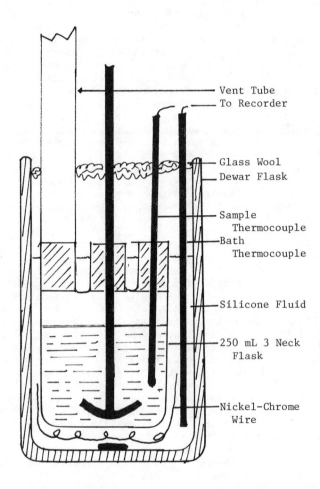

Figure 1 - Diagram of the calorimeter used in the adiabatic
thermal stability studies.

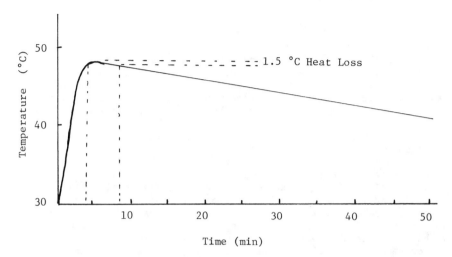

Figure 2 — The experimental differential temperature curve obtained from the adiabatic thermal stability study.

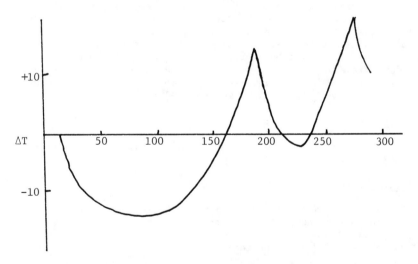

Figure 3 — Temperature curve obtained during the adiabatic temperature rise study.

The heat capacity of the original procedure was 140 cal/mole °C (8). The adiabatic temperature rise was then calculated to be:

$$12.5 \text{ k cal/mole} = (\triangle T)(0.140 \text{ Kcal/mole } °C)$$
$$\triangle T = 89°C$$

The reason the original procedure was not safe was then evident. The original reaction was carried out at a temperature of 90°C, without cooling, and the heat of the reaction drove the reaction temperature into a region of exothermic activity, resulting in a runaway reaction. The greater heat capacity of the revised system, due to the excess sulfuric acid, absorbed most of the heat of the reaction, preventing the reaction temperature from rising to a dangerous level.

In summary, we determined through simple thermodynamic calculations that a potential safety problem did exist, and how to diminish this problem by the addition of excess sulfuric acid. We investigated the exothermic behavior of the revised nitration using common laboratory equipment, and found that an exotherm did occur initiating at 100°C. We determined the adiabatic temperature rise to be less than 20°C which would insure that the nitration reaction, properly batched, would not approach a temperature of exothermic activity. Finally, we explained the reason for the thermal instability of the original procedure.

Of course, the revised nitration reaction is not free of hazards associated with human error or equipment failure. It is important to note that at Parke-Davis, we recognize that the person most familiar with a particular reaction is the development chemist. Therefore, the development chemist actually carries out the reaction in our pilot plant, and he or she can best recognize when a hazardous situation arises. Work is continuing on this reaction to further decrease the risk involved. Preliminary results using one molar equivalent of nitric acid versus CDMP look promising.

Acknowledgment

The author wishes to acknowledge Mr. Charles Combs for his critical discussions during the course of this work.

Literature Cited

1. Fawcett, Howard H., Wood, William S. "Safety and Accident Prevention in Chemical Operation"; Wiley: New York, 1965; Chap. 19.
2. Coffee, R.D. Loss Prevention 1969, 3, 18.
3. Van Dolah, R. W. Loss Prevention 1969, 3, 32.
4. Grelecki, C. National Safety Congress Transactions 1973, 5, 22.
5. Castellan, G. W. "Physical Chemistry Second Edition"; Addison-Wesley: Reading, Mass., 1971, p. 137.
6. Pfeiffer, C., personal communication.
7. De La Mare, P. B. D.; Ridd, J. H., "Aromatic Substitution: Nitration and Halogenation"; Butterworth: London, 1959, Chap. 5.
8. Perry, P. H.; "Chemical Engineers' Handbook"; Magraw-Hill: New York, 1973, p. 3-205.

RECEIVED November 14, 1984

Author Index

Brannegan, Daniel P., 17
Chakrabarti, Ashok, 91
Chase, Malcolm W., 81
Davies, Carole A., 81
DuVal, Robert C., 57
Gibson, S.B., 33
Handley, John R., 41
Hoffmann, John M., 1
Hofmann, Mary J., 7

Kipnis, Irving M., 81
Shafaghi, A., 33
Steiner, Edwin C., 91
Treweek, Dale N., 81
Van Horn, David J., 23
Van Roekel, Linda, 69
Werling, Craig L., 91
Yoshimine, Mas, 91
Zeller, James R., 107

Subject Index

A

Accelerating rate calorimeter (ARC),
    hazard evaluation process, 50
Accuracy, CHETAH program, 86
Addition
    continuous, DACSL kinetic and
        reactor modeling, 97
    instantaneous, simulation, DACSL
        program, 101
Additivity methods
    comparison, CHETAH program, 83t
    group, CHETAH program, 82
Adiabatic temperature rise
    CDMP nitration, 114
    exothermic reaction, 70
Agenda, process hazards review,
    typical, 12-14
American Society for Testing and
    Materials (ASTM), CHETAH
    program, 81
Analytical equipment, purchase
    requiring hazard review, 9
Applications, various, CHETAH
    program, 86
Atmospheres, various, thermal
    instability data, 65t
Avoidance of hazards in
    pharmaceuticals processing, 41-55

B

Batch reactor, typical, DACSL kinetic
    and reactor modeling, 94
Benefits, HAZOP study, 37
Benson's second-order group additivity
    method, CHETAH program, 82
Benzoyl peroxide (BPO), effect of
    dilution with xylene, CHETAH
    program, 87
tert-Butyl hydroperoxide (TBHP),
    effect of dilution with xylene,
    CHETAH program, 87

C

Calorimeter
    accelerating rate, hazard evaluation
        process, 50
    hazard evaluation process, 53f
Calorimetry
    DSC, initial screening for
        thermochemical hazards, 59
    thermal instability data, 62-67
Cautions, investigations of thermal
    hazards, 79

Change analysis
   introduction of new hazards, 19
   simple risk assessment technique, 26
Checklist
   preliminary hazard analysis,
      outline, 29
   types of reviews, 10
Chemical process hazard review
   overview, 1-6
   potential thermochemical, 57
Chemical reaction energy
   categorization, hazard
      potential, 3t
Chemical research environment, process
   hazard review, 7-15
Chemical thermodynamic and energy
   release evaluation (CHETAH)
      estimation of thermochemical and
         hazard data, 81-90
      new, overview, 89
Chemist, hazard assessment
   responsibility, 46
Chemistry
   exothermic reaction,
      discussion, 69-73
   process, and hazard review, 3
5-Chloro-1,3-dimethyl-1H-pyrazole,
   nitration risk-assessment before
      scale-up, 107-14
Classification, compound, CHETAH
   program, 86
Combustion, heat of, CHETAH
   program, 85
Committee, process hazards review,
   development, 12
Condensed phase reaction, CHETAH
   program, 88
Condenser, reflux conditions, DACSL
   kinetic and reactor modeling, 97
Contamination in production, potential
   hazard sources, 42
Critical incident technique, risk
   assessment, 25
Critical parameters, Semenov theory,
   thermal hazard evaluation, 76

                              D

Decomposition
   heat of, thermal instability
      data, 65t
   maximum energy of, CHETAH
      program, 85

Definitions
   DACSL kinetic and reactor
      modeling, 104
   HAZOP guide words, 36
   HAZOP study, 27
   PHR, 8
Detonation potential, CDMP
   nitration, 108
Diazonium fluoroborate salt, organic,
   thermolysis, 41
Differential equations, DACSL kinetic
   and reactor modeling, 94
Differential scanning calorimetry
   (DSC)
   curves, $\underline{N},\underline{N}$-diphenylhydrazine
      hydrochloride, 61f
   initial screening, thermochemical
      hazards, 59
Differential thermal analysis (DTA)
   initial screening, thermochemical
      hazards, 59
   recorder traces, time-dependent
      thermal stability, 61f
Dilution
   CHETAH program simulation, 87
   effect on CDMP nitration, 109
$\underline{N},\underline{N}$-Diphenylhydrazine hydrochloride
   DSC curves, 61f
   DSC data, 59
Disposal, reaction hazard
   evaluation, 19
Dow advanced continuous simulation
   language (DACSL), kinetic and
      reactor modeling, 91-105
Du Pont Experimental Station, process
   hazard review program, 7-15
Dynamic heating methods
   initial screening, thermochemical
      hazards, 59
   thermal instability data, 60

                              E

Elements, new CHETAH program, 89
Energy categorization, chemical
   reaction hazard potential, 3t
Energy hazard evaluation, CHETAH
   program, 85
Energy of decomposition, maximum,
   CHETAH program, 85
Energy release appraisal, TNT, CHETAH
   program, 87t
Engineer, process, hazard assessment
   responsibility, 46

Engineering
    considerations, hazard evaluation
        process, 50
    large-scale system, thermal runaway
        hazard evaluation, 74
Enthalpy change, exothermic
    reaction, 70
Enthalpy of formation, CHETAH
    program, 85
Enthalpy term, DACSL kinetic and
    reactor modeling, 96
Environmental, safety, health review
    procedure, outline, 30
Environmental coordinator, hazard
    assessment responsibility, 46
Environmental review, safety and
    health, risk assessment
    technique, 26
Equipment, processing, prestart-up
    review, 15
Ester, α-oximino intermediate, thermal
    stability hazard, 67t
Evaluation
    hazards in CDMP nitration, 107-14
    hazards in process
        development, 17-21
    hazards in thermal runaway
        reaction, 69-79
    initial hazard, objectives, 42-45
    thermochemical hazard, 57-68
Exothermic apparatus, hazard
    evaluation process, 52f
Exothermic reaction
    CDMP nitration, 110
    chemistry, discussion, 69-73
    potential hazard sources, 42
    temperature rise factors, 111

F

Failure mode and effect analysis (FM&E)
    simple risk assessment technique, 25
    types of reviews, 10
Fault tree analysis
    risk-assessment technique, 27
    types of reviews, 11
Flexibility, HAZOP method, 39
Flow chart, hazard review process, 46
Flow reactor, PHR, 14
Flowrate, fluid through the jacket,
    DACSL kinetic and reactor
    modeling, 97
Fluid
    flowrate through the jacket, DACSL
        kinetic and reactor modeling, 97
    jacket, loss, simulation, DACSL
        program, 101

Fluoroborate salts, unexpected
    decomposition, 42
Formation, enthalpy of, CHETAH
    program, 85
Frequency, process hazards review, 14

G

Gas
    evolved, thermal instability
        data, 65t
    rate of accumulation in head space,
        DACSL kinetic and reactor
        modeling, 96
Gas phase calculations, CHETAH
    program, 81
Generic controls, unknown hazards, 19
Group additivity methods, CHETAH
    program, 82
Guide words
    application, HAZOP, 35
    definitions, HAZOP, 36
Guidelines, when to conduct a PHR, 9

H

Hazard(s)
    evaluation
        CDMP nitration, 107-14
        primary responsibility, 18
        process development, 17-21
    factors that produce, 23
    introduction of new, 19
    potential, identification, 33
    thermal, cautions in
        investigations of, 79
Hazard analysis, preliminary
    checklist, 29
    discussion, 24
Hazard and operability study (HAZOP)
    discussion, 33-39
    questionnaire, 49
    risk assessment technique, 27
    types of reviews, 10
Hazard avoidance, pharmaceuticals
    processing, 41-55
Hazard data, estimation, CHETAH
    program, 81-90
Hazard evaluation
    chemist, hazard assessment
        responsibility, 46
    complex reaction, kinetic and
        reactor modeling, 91-105
    energy, CHETAH program, 85

Hazard evaluation--Continued
  form, 46,48f
  initial objectives, 42-45
  responsibility, 45
  thermal runaway reaction, 69-79
  thermochemical, 57-68
Hazard review
  chemical process, overview, 1-6
  sequence, 47f
  summary, hazard evaluation
    process, 54
  team
    organization, 45

  process examination, 46-49
Health considerations, hazard
  evaluation process, 50
Health, safety, and environmental review
  procedure outline, 30
  risk assessment technique, 26
Heat capacity, exothermic reaction, 71
Heat of combustion, CHETAH program, 85
Heat of decomposition, thermal
  instability data, 65t
Heat of reaction
  calculations, CHETAH program, 84
  exothermic reaction, 70
Heat production rate, exothermic
  reaction, 71
Heating methods, various, thermal
  instability data, 60
Heterogeneous catalysts, PHR, 14
Hot spots, potential hazard
  sources, 42

                      I

Incident recall, simple risk
  assessment technique, 25
Information gathering techniques, 25
Instantaneous addition, simulation,
  DACSL program, 101
Isothermal heating methods, thermal
  instability data, 60
Isothermal laboratory reactors, DACSL
  kinetic and reactor modeling, 93

                      J

Jacket control parameters, DACSL
  kinetic and reactor modeling, 97
Jacket fluid, loss, simulation, DACSL
  program, 101

Jacket inlet temperature, DACSL
  kinetic and reactor
  modeling, 94,96
Job safety analysis, simple risk
  assessment technique, 26

                      K

Kinetic approximations, thermal
  instability data, 65t
Kinetic modeling, hazard evaluation
  and scale-up of a complex
  reaction, 91-105
Kinetic parameters, exothermic
  reaction, 70

                      L

Laboratory operations, requiring a
  hazard review, 9
Large-scale system, engineering,
  thermal runaway hazard
  evaluation, 70
Literature review
  hazard review process, 4
  pharmaceutical hazard review, 58

                      M

Management oversight and risk tree
  (MORT), risk assessment
  technique, 28
Mass transfer effect, simple,
  potential hazard sources, 42
Material
  added, DACSL kinetic and reactor
    modeling, 97
  testing and hazard review process, 5

                      N

New product, hazard assessment
  responsibility, 45
Nitration, CDMP, risk assessment
  before scale-up, 107-14

O

Open test tube thermal stability test,
    CDMP nitration, 110t
Optimization, DACSL kinetic and
    reactor modeling, 97
Organization, hazard review team, 45
OSHA Hazards Communications Standard,
    effect on hazard evaluation, 21
α-Oximino ester intermediate, thermal
    stability hazard, 67t
Oxy-Claisen rearrangement, CHETAH
    program, 89
Oxygen, various amounts, effect on
    CDMP nitration, 108
Oxygen balance, CHETAH program, 86

P

Parameters, various, to define
    exothermic reaction, 70t
Parameter estimations, DACSL kinetic
    and reactor modeling, 94
Peroxides, effect of dilution with
    xylene, CHETAH program, 87
Pharmaceuticals
    evaluation of thermochemical
        hazards, 57
    hazard avoidance, 41–55
Phases, various, CHETAH program, 88
Phosphorus oxychloride, quenching,
    potential hazards, 19
Physical parameters
    exothermic reaction, 71
    influence on test data, 5
Physical testing
    hazard evaluation process, 50
    limitations, thermochemical hazard
        review, 67
    thermally unstable materials, 58
Pilot plant director, hazard
    assessment responsibility, 45
Pilot plant scale-up, CDMP nitration
    risk assessment, 107–14
Plant safety officer, hazard
    assessment responsibility, 46
Potential hazards, identification, 33
Preliminary hazard analysis (PHA)
    checklist, 29
    discussion, 24
Pressure
    effect on thermal instability
        data, 60,65t
    generation rate, exothermic
        reaction, 71

Prestart-up review/process hazards
    audit, definitions, 8
Priority, developing a safe
    process, 24
Procedure, review, safety, health and
    environmental, outline, 30
Process contributions, HAZOP study, 37
Process development
    applicability of HAZOP method, 39
    goal, 17
    hazards evaluation, 17–21
    pharmaceuticals, hazard
        avoidance, 41–55
Process engineer, hazard assessment
    responsibility, 46
Process equipment, prestart-up
    review, 15
Process examination, hazard review
    team, 46–49
Process hazards review (PHR)
    definitions, 8
    objectives, 2
Process supervisor, primary
    responsibility for hazards
    evaluation, 18
Project contributions, HAZOP study, 37
Pump, liquid, addition to flow
    reactor, PHR, 14

Q

Quenching, reaction hazard
    evaluation, 19

R

Radius, critical, Semenov theory,
    thermal hazard evaluation, 76
Reaction
    thermal runaway, hazard
        evaluation, 69–79
    various, DACSL kinetic and reactor
        modeling, 92f
Reaction calorimetry, thermal
    instability data, 65–67
Reaction conditions
    identification and quantification, 3
    simulation, DACSL kinetic and
        reactor modeling, 97
Reaction energy categorization,
    chemical, hazard potential, 3t
Reaction rates, DACSL kinetic and
    reactor modeling, 94

Reactor modeling, hazard evaluation
  and scale-up of a complex
  reaction, 91–105
Reflux conditions in the condenser,
  DACSL kinetic and reactor
  modeling, 97
Rensselaer plant, Sterling Organics,
  hazard avoidance, 41–55
Research and development chemist,
  hazard assessment
  responsibility, 46
Research environment, chemical process
  hazard review, 7–15
Review
  chemical process hazard,
    overview, 1–6
  hazard, various types, 10
  safety, health, and environmental,
    outline, 30
  selection method, 11
Risk assessment
  CDMP nitration, 107–14
  techniques for
    experimentalists, 23–31
Risk identification, systematic,
  techniques, 25
Risk-reducing procedures, accumulation
  during process development, 18
Risk tree, management oversight and,
  risk assessment technique, 28
Runaway reaction
  CDMP nitration, 114
  simulation, DACSL program, 98

                           S

Safety analysis
  job, simple risk assessment
    technique, 26
  system, HAZOP study, 34
Safety information, accumulation
  during process development, 18
Safety officer, plant, hazard
  assessment responsibility, 46
Safety problem, potential, CDMP
  nitration, 114
Safety study evaluation, HAZOP
  study, 36
Safety, health, environmental
  review (SHE)
  outline, 30
  risk assessment technique, 26
Scale-up
  CDMP nitration risk
    assessment, 107–14
  CHETAH program, 87

Scale-up--Continued
  complex reaction, kinetic and
    reactor modeling, 91–105
  introduction of new hazards, 19
  potential hazard sources, 42
Schiemann reaction, unexpected
  hazard, 41
Screening
  initial, thermochemical hazards, 59
  potential hazards, 20
Scrubbing, reaction hazard
  evaluation, 19
Semenov theory
  cautions in investigations of
    thermal hazards, 79
  thermal runaway hazard
    evaluation, 74–79
Solvent, effect, thermal instability
  data, 65t
Stability
  thermal
    open test tube, CDMP
      nitration, 110t
    pharmaceutical hazard review, 58
  time-dependent thermal, DTA recorder
    traces, 61f
Stainless steel, effect on thermal
  instability data, 62
Sterling Organics, Rensselaer plant,
  hazard avoidance, 41–55
Study planning, HAZOP, 35
Substitutions, CHETAH program, 84
Sulfuric acid, effect on CDMP
  nitration, 109
Systems
  applicability of HAZOP method, 39
  safety analysis, HAZOP study, 34

                           T

Team
  assembly, HAZOP, 35
  hazard review, process
    examination, 46–49
  organization, hazard review, 45
Temperature
  adiabatic, CDMP nitration, 114
  CDMP nitration, 109
  critical, Semenov theory, thermal
    hazard evaluation, 76
  effect on thermal instability
    data, 60
  jacket inlet, DACSL kinetic and
    reactor modeling, 94,96
  range, CHETAH program, 81
  rise, adiabatic, exothermic
    reaction, 70

Temperature--Continued
  set, DACSL kinetic and reactor
    modeling, 97
  stepping simulation, DACSL
    program, 98
Testing
  DACSL kinetic and reactor
    modeling, 93
  hazard review process, 5
  screening for potential hazards, 20
Theoretical calculations, and hazard
  review process, 4
Thermal conductivity, exothermic
  reaction, 71
Thermal hazards, cautions in
  investigations of, 79
Thermal instability
  CDMP nitration, 109
  follow-up testing, 62
  physical testing program, 58
Thermal runaway reaction, hazard
  evaluation, 69-79
Thermal stability
  hazard, α-oximino ester
    intermediate, 67t
  open test tube, CDMP nitration, 110t
  testing, pharmaceutical hazard
    review, 58
  time-dependent, DTA recorder
    traces, 61f
Thermochemical calculations, CHETAH
  program, 84
Thermochemical data, estimation,
  CHETAH program, 81-90
Thermochemical hazard
  evaluation, 57-68

Thermodynamic calculations, CDMP
  nitration, 114
Thermodynamic data, CHETAH
  program, 85t
Thermodynamic parameters, exothermic
  reaction, 70
Thermolysis, organic diazonium
  fluoroborate salt, 41
Time-dependent thermal stability, DTA
  recorder traces, 61f
Time segments, DACSL kinetic and
  reactor modeling, 97
Transesterification, CHETAH
  program, 88
2,4,6-Trinitrotoluene (TNT), energy release
  appraisal, CHETAH program, 87t

V

Vaporizer, addition to flow reactor,
  PHR, 14
Vent size, various, simulation, DACSL
  program, 101
Volume, critical, Semenov theory,
  thermal hazard evaluation, 76

W

What if review--See Failure mode and
  effect analysis
Wilcox-Bromley method, new CHETAH
  program, 89

Production and indexing by Susan Robinson
Jacket design by Pamela Lewis

Elements typeset by Hot Type Ltd., Washington, D.C.
Printed and bound by Maple Press Co., York, Pa.

RECENT ACS BOOKS

"Purification of Fermentation Products:
Applications to Large-Scale Processes"
Edited by Derek LeRoith, Joseph Shiloach, and Timothy J. Leahy
ACS SYMPOSIUM SERIES 271; 200 pp.; ISBN 0-8412-0890-5

"Reaction Injection Molding: Polymer Chemistry and Engineering"
Edited by Jiri E. Kresta
ACS SYMPOSIUM SERIES 270; 302 pp.; ISBN 0-8412-0888-3

"Materials Science of Synthetic Membranes"
Edited by Douglas R. Lloyd
ACS SYMPOSIUM SERIES 269; 496 pp.; ISBN 0-8412-0887-5

"The Chemistry of Allelopathy: Biochemical Interactions Among Plants"
Edited by A. C. Thompson
ACS SYMPOSIUM SERIES 268; 466 pp.; ISBN 0-8412-0886-7

"Environmental Sampling for Hazardous Wastes"
Edited by Glenn E. Schweitzer and John A. Santolucito
ACS SYMPOSIUM SERIES 267; 144 pp.; ISBN 0-8412-0884-0

"Materials for Microlithography: Radiation-Sensitive Polymers"
Edited by L. F. Thompson, C. G. Willson, and J. M. J. Frechet
ACS SYMPOSIUM SERIES 266; 490 pp.; ISBN 0-8412-0871-9

"Computers in the Laboratory"
Edited by Joseph Liscouski
ACS SYMPOSIUM SERIES 265; 136 pp.; ISBN 0-8412-0867-0

"The Chemistry of Low-Rank Coals"
Edited by Harold H. Schobert
ACS SYMPOSIUM SERIES 264; 328 pp.; ISBN 0-8412-0866-2

"Resonances in Electron-Molecule Scattering, van der
Waals Complexes, and Reactive Chemical Dynamics"
Edited by Donald G. Truhlar
ACS SYMPOSIUM SERIES 263; 536 pp.; ISBN 0-8412-0865-4

"Seafood Toxins"
Edited by Edward P. Ragelis
ACS SYMPOSIUM SERIES 262; 473 pp.; ISBN 0-8412-0863-8

"Computers in Flavor and Fragrance Research"
Edited by Craig B. Warren and John P. Walradt
ACS SYMPOSIUM SERIES 261; 157 pp.; ISBN 0-8412-0861-1

"Rubber-Modified Thermoset Resins"
Edited by Keith Riew and J. K. Gillham
ADVANCES IN CHEMISTRY SERIES 208; 370 pp.; ISBN 0-8412-0828-X

"The Chemistry of Solid Wood"
Edited by Roger M. Rowell
ADVANCES IN CHEMISTRY SERIES 207; 588 pp.; ISBN 0-8412-0796-8